W9-CDT-757

A Sheltered Life

Also by Paul Chambers

Bones of Contention

A Sheltered Life

The Unexpected History of the Giant Tortoise

PAUL CHAMBERS

BOWLING GREEN STATE
UNIVERSITY LIBRARIES

OXFORD
UNIVERSITY PRESS

OXFORD
UNIVERSITY PRESS

Oxford University Press, Inc., publishes works that further
Oxford University's objective of excellence
in research, scholarship, and education.

Oxford   New York
Auckland   Cape Town   Dar es Salaam   Hong Kong   Karachi
Kuala Lumpur   Madrid   Melbourne   Mexico City   Nairobi
New Delhi   Shanghai   Taipei   Toronto

With offices in
Argentina   Austria   Brazil   Chile   Czech Republic   France   Greece
Guatemala   Hungary   Italy   Japan   Poland   Portugal   Singapore
South Korea   Switzerland   Thailand   Turkey   Ukraine   Vietnam

Copyright © 2006 by Paul Chambers

First published in Great Britain in 2004 by John Murray (Publishers)
A division of Hodder Headline

The right of Paul Chambers to be indentified as the Author of the Work
has been asserted by him in accordance with

the Copyright, Designs and Patents Act 1988.

Published by Oxford University Press, Inc. in 2006
198 Madison Avenue, New York, New York 10016

www.oup.com

Oxford is a registered trademark of Oxford University Press

All rights reserved. No part of this publication may be reproduced,
stored in a retrieval system, or transmitted, in any form or by any means,
electronic, mechanical, photocopying, recording, or otherwise,
without the prior permission of Oxford University Press.

Library of Congress Cataloging-in-Publication Data
Chambers, Paul, 1968–
A sheltered life : the unexpected history of the giant tortoise / Paul Chambers.
p. cm.
Originally published: London : John Murray, © 2004
Includes bibliographical references and index.
ISBN-13: 978-0-19-522396-5
ISBN: 0-19-522396-9
1. Testudinidae. I. Title.
QL666.C584.C46 2005
597.92'46—dc22   2005050859

1 3 5 7 9 8 6 4 2

Printed in the United States of America
on acid-free paper

For my parents, Martyn and Sally Chambers

Contents

CONTENTS

PART FOUR Obsession

PART FIVE Pets

PART SIX Recovery

PART SEVEN In the Blood

Illustrations

Acknowledgements

The researching and writing of this book was by no means a solo process and my profound thanks go to all those people who were kind enough to offer me their time, advice and knowledge. In particular I would like to single out the following for their invaluable assistance: Scott Davis, Alex Freeman, Carol Gokce, Tim Haines, Jasper James, Colin McCarthy, Adam Perkins, Kate Pickard, Frank Sulloway, Polly Tucker, Scott Thomson and Vern Weitzel. I would also like to thank the staff at the following institutions for helping me track down obscure books, papers and original archive material: Natural History Museum (London), University College London, Charles Darwin Foundation (London), Society of Genealogists (London), Cambridge University Library (Cambridge, England), Down House (Kent, England), John Oxley Library (Brisbane, Australia), Sydney Library (Australia), Hertfordshire Library Service (England), National Army Museum (London), Australia Zoo (Queensland), British Newspaper Library (London), Kew Gardens (London), Institute of Historical Research (London) and Bodleian Library (Oxford, England).

A big thank-you must go to my agent, Sugra Zaman of Watson Little Ltd, who has worked tirelessly on my behalf over the years, and also the magnificent staff at John Murray Ltd, especially Grant McIntyre, Lizzie Dipple and John Murray, whose advice and patience is greatly appreciated. I must add a special thank-you to John Baxter for his candid and excellent advice concerning the draft manuscript.

Finally, I must thank my wife Rachel and my parents and parents-in-law for their unceasing encouragement and unending sympathy.

Thank you, one and all.

Scientific and Other Conventions

In writing this book my aim is to tell the story of the world's giant tortoises and their relationship with humanity. I particularly did not wish to create a detailed scientific textbook on tortoise biology. As such I have generally tried to keep the text free of complicated terms, complex techniques and overlong discussions. Also, because of pressure of space and the need to keep the story flowing, I have tended to focus on the key personalities and events in the tortoises' story, which means that some of the minutiae have been left out. Those who wish to delve more deeply into the events and issues described and discussed here should refer to the Notes and Sources, p. 275.

I have also taken the liberty of updating most of the old-fashioned spelling used in many of the quotes, and have used the modern Spanish names for the Galápagos rather than the traditional English ones.

Preface

Santa Maria Island, Galápagos Archipelago
25 September 1835

The party of men paused to catch their breath. They were weary from the vigorous climb up a rough and dusty path that took an unmerciful line straight up the side of a steep volcano.

'Of course, you're seeing this island at its very best,' opined the middle-aged man standing at the head of the group. 'We've actually had some rainfall to speak of this year. Sometimes we get almost nothing at all.'

His companions listened with incredulity, some looking back down the hill towards the natural harbour from whence they had set off earlier that morning. As it had others who had trod this path before them, the idea that their host believed this to be a well-watered landscape made them dread to think what it must look like under drought conditions.

For much of the climb the landscape on either side of the path showed every sign of being parched to within an inch of its life. The trees, if they could be called that, were stunted and almost leafless, their roots fighting to burrow into the thin volcanic soil. Underneath these were patches of brown grass whose fronds appeared brittle and faded in the bright mid-morning sunshine. There was very little sign that any of these plants had received any rain during the last six years, let alone the last six weeks.

After a short rest the men continued on their journey, following the path around one side of a high hill. A cooling southerly breeze greeted them as they left the shelter of the dry scrubland and started to walk across a flat and open plateau on which several dozen primitive wooden thatched houses had been erected. 'Welcome,' said the party's leader, 'to the settlement.'

The young men found themselves looking at a verdant scene of green grass, banana trees and healthy-looking crops. After their weeks of living at sea, the colours, together with the sight of people, dogs, goats and pigs, not to mention the cacophony of associated noise, all served temporarily to swamp their senses.

Given that it was stuck halfway up the side of a volcano that was itself in one of the most remote locations on earth, the settlement was not the sort of place that most sane people would aspire to, but to sailors who had spent weeks aboard a small ship staring at nothing but the sea and sky it looked remarkably civilised.

The middle-aged man led his companions across the plateau and further up the hill to his own house. This was larger than the rest, better furnished and within reach of the permanent freshwater spring on which the settlement was founded. It was, he explained, one of the few privileges of being the official governor in such a remote location. Bidding the men welcome and, after seating them in the shade, he walked over to speak to some of the villagers who, curious about the visitors, had wandered up to the house.

The visitors began to relax, opening the bottle of rum that they had taken with them from the ship, and basking in the warm sun. As the governor returned to his guests, a broad smile spread across his face.

'You will, of course, stay to lunch?' he said hopefully. 'I can offer you goat, pig, or tortoise.'

PART ONE

Discovery

A Sheltered Life

O UR PLANET IS no stranger to giant animals. There were, of course, the dinosaurs of the Mesozoic era and, in more recent times, woolly mammoths, giant deer, cave lions and even giant killer kangaroos.

For over half a billion years, being big has been used as a survival strategy by a wide range of animals but about 2 million years ago, at the start of the ice age, there was a change in fortune for the world's giants. The swift ecological shifts that accompanied the advancing and retreating glaciers were hard for the larger animals to adapt to, forcing some of them into a terminal decline. Other species, such as the mammoth, were possibly hunted to extinction by humans. Whatever the cause, the giants' demise was so rapid that by the time the Egyptians were building their first pyramids the largest land animal was the African elephant. The giant sloths, mammoths, killer kangaroos and the like had all joined the long list of species that have been forced to leave the earth's genepool.

During this phase of gigantism it was not just mammals that became big; some reptiles used the same strategy. Fossils of large snakes, lizards and crocodiles have been found across the world but the most widely found giant reptile was the giant tortoise whose remains have been recovered from places as diverse as Malta, India, Brazil, Madagascar and Indonesia. Some of these were truly gargantuan, with shell lengths measuring up to two metres and a live weight of 500 kilograms or more, but even they did not escape the ice age extinctions.

One by one each species of giant tortoise vanished from the earth, leaving behind only their bones for palaeontologists to puzzle over. By the end of the last ice age, 10,000 years ago, the mainland regions of the world were devoid of giant tortoises. By rights this species should have become entirely extinct but through a stroke of luck some of their number had stumbled into the sea and were carried by ocean currents to remote islands, far from major continents.

Once stranded on these islands, the giant tortoises found themselves surrounded by an abundance of vegetation to feed on but with none of the mainland predators that had previously troubled them. Under these conditions they could live and breed without hindrance, their numbers multiplying and their size increasing, which enabled their bodies to store water and fat, helping them through the lean times.

While the giant tortoises on the mainland succumbed to the wave of extinction that affected (and continues to affect) so many of the world's large animals, the island tortoises were spared. They were now the largest reptiles on earth and the sole survivors of a race of giant tortoises that had once spread across the planet. For hundreds of thousands of years they led a sheltered island life, oblivious to the changes that were taking place in the world around them. While the tortoises continued to flourish in their island paradises, another of the earth's successful post-ice age species, *Homo sapiens*, had learnt the art of toolmaking and was seeking to impose its will upon the entire planet.

～

The sixteenth century was a time of reconnaissance for the European seafaring nations. Each year hundreds of well-equipped but fragile ships would leave their home ports on long voyages that took them either west towards the New World or east towards the East Indies.

Navigation was then only a rudimentary science and sea charts were notoriously unreliable. In many cases getting a ship to its destination was as much a matter of luck as skill. The inevitable mistakes in navigation sent ships far off course and caused them to stumble randomly upon remote and uninhabited islands. Some of these proved to be of great economic and political significance; others were thought so worthless that their discoverers did not even bother to mark their location on a map. In 1497, Portuguese sailing ships rounded the Cape of Good Hope for the first time and began to ply across the Indian Ocean. Their goal was to find a route to the Malabar

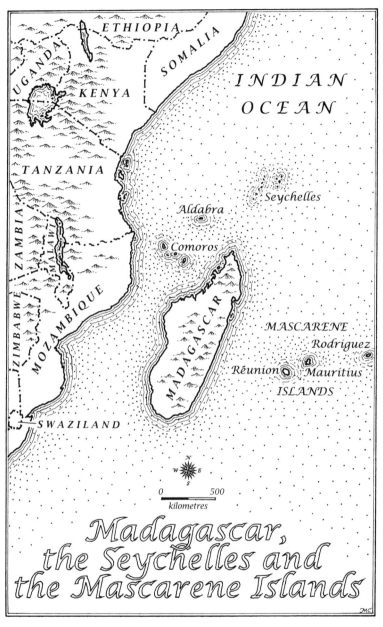

A map of the Indian Ocean and its giant tortoise bearing islands.

coast of India and thus gain access to its spices and mineral wealth. They were successful and within a few years there was a small Portuguese colony established along the Calcutta coast to which several dozen ships would annually make the hazardous journey. It was thanks to a navigation error during one such voyage to India that, in 1505, the Portuguese first sighted the huge island of Madagascar, which they initially named St Laurentio.

A few months later, further navigation errors led Don Pedro Mascaregnas to discover two further islands lying several hundred kilometres to the east of Madagascar. He gave the larger one the name Cerné and the smaller one he named Mascaregnas after himself. We now know these islands respectively as Mauritius and Réunion although they are still collectively called the Mascarene Islands after their discoverer. Having discovered and named the two Mascarene Islands, the Portuguese seem to have had little interest in them and, after a brief visit, left them alone.

Despite now adorning the primitive nautical maps of the Indian Ocean, Mauritius and Réunion were ignored for the remainder of the sixteenth century and received few visitors. In 1598 the islands nominally changed their nationality after being accidentally rediscovered by a Dutch flotilla. Cerné island, as it was still known, was renamed Mauritius in honour of the Dutch king and, for the first time, detailed notes were made of its bays and harbours. At the same time the Dutch also laid claim to Réunion but chose not to colonise either island. Instead they released a few chickens into the wild, weighed anchor and moved on. It was not until 1638 that the Mascarene group gained its third and final member when the small island of Rodriguez, 400 kilometres to the east of Réunion, was discovered, again by the Dutch.

By the early seventeenth century, the sea routes to India had become important to the Portuguese, the Dutch, the French and the English. Inevitably, tension grew between these various nations, and governments began to look for suitable harbours from which naval attacks could be launched on enemy vessels. Being directly on the route between India and South Africa, the islands of Mauritius, Réunion and Rodriguez were ideal candidates for naval bases. It was now only a matter of time before the Mascarene Islands received their first full-time colonists.

For some years before colonisation, ships had been stopping at the Mascarene Islands for food and water. Some of them returned with reports of the strange wildlife there. Stories were told of giant birds that ate stones and were so unafraid of humans that they could be caught by hand and strangled. Strangest of all were the tales of huge tortoises whose size and strength was such that they could carry several men on their backs and still walk with comfort.

'Here,' wrote an anonymous Dutch visitor to Mauritius in 1598, 'are turtles of such large dimensions that one of their shells was sufficiently capacious to admit of six men to take their seats in it.'

As the number of visiting ships increased, so the native wildlife of the Mascarene Islands was brought into ever closer contact with humanity. After several hundred thousand years of living a sheltered life, the giant tortoises were about to make contact with the wider world and the end result was not going to be pleasant.

The Worthless Islands

T HE DISCOVERY OF new islands populated by strange animals was not confined to the Indian Ocean. At the same time that the giant tortoises of the Mascarenes were coming to the attention of Portuguese sailors, other European explorers were making discoveries of their own on the opposite side of the world.

The year was 1535 and for some time conquistadors had been rampaging through South America in search of land and gold. Such ventures were far from chivalrous and in Peru a minor spat broke out between two of the Spanish king's favourite generals. A peacemaker was needed and in the absence of any other candidate the job fell to Tomás de Berlanga, the bishop of Panama. The reluctant Bishop de Berlanga was given a ship and ordered to sail to Peru in order to quell dissent in the new colonies.

Berlanga set sail from Panama on 23 February and for several days made good progress as his ship was borne southwards on a strong breeze. The South American coastline was still uncharted but by hugging its bays and coves and travelling in a southerly direction, Berlanga knew that he could not fail to find his ultimate destination, the Peruvian town of Puerto Viejo. The journey should have been straightforward, if somewhat tedious, but luck was not on Berlanga's side.

On the eighth day at sea Berlanga's ship entered the Doldrums, an area of slack wind on either side of the equator. The ship was becalmed, her magnificent sails hanging loose in the fetid aid, but while the winds were still the waters beneath most certainly were not. Unbeknown to Berlanga and his crew, his vessel had strayed into a strong and swift-flowing current that was moving away from the South American coast and heading directly westwards into the heart of the Pacific Ocean.

The first inkling that the sailors had of their being caught in a strong current came when the South American coastline, which they had so carefully kept in sight for over a week, began to grow smaller and more distant until eventually it disappeared altogether. The boat was now like a swimmer caught in a rip tide and any attempt by the crew to steer or control their vessel proved fruitless. Until the winds returned, the crew were at the mercy of the sea. All they could do was wait and hope that they would not be dragged too far from the shore.

A becalmed ship out of the reach of land presents a great hazard to those on board. In Panama, Berlanga had loaded enough supplies to feed and water his crew and their horses for about two weeks, which would have been sufficient to get him about three-quarters of the way to his destination.

It is therefore not surprising that by their fourth day at the mercy of the current, and with still no sight of land, a sense of panic began to grow among officers and crew alike. By Berlanga's own reckoning, they were down to their last two days' worth of water. Even the most simple-minded sailor could work out that if it took them another six days to find land again (bearing in mind that in those days there was no means of calculating a longitudinal position), a good portion of the crew would be dead on arrival. The idea of a tormented death from thirst was a real fear for mariners. It was not unknown for men to throw themselves into the sea to be drowned or eaten by sharks rather than to die by inches, slowly dehydrating in the hot sun. If life was hard for Berlanga's human crew, then things were even worse for the horses, whose water supplies had been withdrawn some days beforehand.

It was on day six, when the situation was already critical, that the crew believed their luck to be changing. There, on the horizon, was the unmistakable outline of a small island. What was initially just a hazy dot in the distance gradually became larger and larger as, for once, the currents carried the ship in the right direction. Berlanga and his crew must have thought that their prayers had been answered. Land normally means fresh water. As the island drew within reach, the ship's longboat was prepared and lowered over the side with a company of oarsmen. Berlanga instructed them to return with as much water as they could carry, as well as some grass for the horses.

After several hours, those left on board were relieved to see the longboat returning from the beach but the news it brought was both grim and perplexing. A lengthy search had revealed nothing but vistas of bare, black rock, which formed themselves into a tortured landscape of raised blocks and shallow crevices. There was no trace of either water or edible plants.

With water supplies on board all but exhausted, this was news that could well have marked the crew's doom. Still becalmed, Berlanga let the currents carry his boat further away from the parched island until it was just a speck on the horizon. The morale of the crew was now at rock bottom. Several horses were dead and many of the crew were so severely dehydrated that they were incapable of performing their duties.

Despite the low morale, the next day brought more hope. Another island was sighted. This time it was large and mountainous and, even from a distance, looked a far more promising bet than the previous day's. Berlanga himself wrote that 'on account of its size and monstrous shape, there [cannot] fail to be rivers and fruits'. With their water supply now gone, all on board knew that this was absolutely their last hope but they were still becalmed and powerless.

For three waterless days the distressed sailors had to watch as the mountainous crags of the island drew slowly nearer, their distant shape offering a vague promise of salvation. The ship moved little by little towards the island, drifting at a snail's pace until, finally, it was within range of one of the many bays around the coast.

The order was given for those who were physically able to land and begin the hunt for water, plus suitable food for the remaining horses. This time Berlanga himself was to join in the search.

He was shocked at what he found. The report given previously by his crew applied equally to this island. Despite its mountainous nature, it seemed to be nothing but bare rock and dry, scrubby plants with not a sign of water anywhere. Rather than despair, Berlanga ordered his crew to dig a well, convinced that the water on the island must lie beneath the surface. This they did and water rushed to fill the freshly dug hole, but on tasting it Berlanga declared that it was 'saltier than sea water' and thus quite useless for drinking purposes. It seemed as though they were again being mocked. With the situation now desperate, Berlanga grouped the men into twos and threes and ordered them to walk into the island's interior to recover whatever water and food they could. Each night they should meet to discuss their progress and future tactics.

At the end of the first day the news was not good. The men had found no water and the only edible vegetation was the fleshy leaves of a strange cactus-like plant. The men 'began to eat of them, and squeeze them to draw all the water from them, and drawn, it looked like slops of lye, and they drank it as if it were rose water'. They had water, of a sort, but the relief was only temporary.

Berlanga had already lost one man and ten horses to thirst. If they were successfully to journey back to the South American mainland, a few fleshy leaves were not going to get them very far; they would need many litres of water. Without it the sailors would remain stranded on the island, forced to scratch out a living until the sun, disease, thirst or hunger claimed them one by one.

The next day Berlanga again ordered his crew out into the mountains and that evening one of the search parties returned with news of a ravine, at the bottom of which was a small but reliable supply of fresh water. There was instantly a rush by every man onshore to reach the spring. Their thirst quenched, the crew then filled every jug, skin and whatever other water-carrying equipment they could muster and took the vital liquid back for their crewmates and the remaining horses. With a reliable supply of water on hand, the crew could at last replenish their supplies and consider making the long journey back to the South American mainland.

Getting water from the ravine and back to the ship was not an easy task. A moderate wind had begun to blow, ending weeks of windlessness. Berlanga was anxious to take advantage of the situation and make a start on the journey home. Being unable to calculate their latitude, he estimated that the islands lay only 'twenty or thirty leagues' (between 135 and 200 kilometres) from South America and that with a moderate breeze the journey could be covered in only a few days.

After they had filled just eight hogsheads with fresh water (each of which held about 240 litres), the order was given to weigh anchor and to steer a course eastwards until land was sighted.

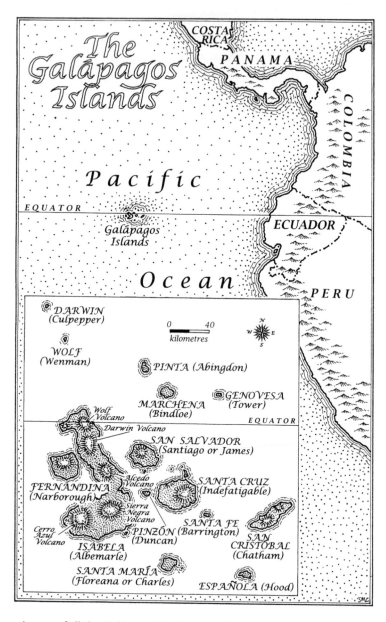

A map of all the Galápagos Islands, several of which are home to giant tortoise populations.

Berlanga's estimation of their distance from South America was a serious error, for during the days when the ship had been becalmed, it had been carried nearly 1,000 kilometres from the coast. Furthermore, to return to South America meant battling the currents and the prevailing winds, both of which were pulling the vessel south and thus increasing their distance from the mainland. After a further eleven days at sea, the crew and horses were again on water rations, with only one full hogshead of water remaining. It was to be a further eight days before land was sighted, by which time the crew were surviving on wine, a feat that must have left them feeling not only very drunk, but also severely dehydrated.

On 9 April the ship entered the Bay of Caraques, where there was a small Spanish colonial outpost and, more importantly, an abundant supply of fresh water. It had been several weeks since the ship's trials had first begun and both men and animals were utterly exhausted.

For Berlanga, the hardships of the voyage had taken their toll. After quelling the Peruvian dispute, he quit the New World and returned home to Spain for a well-earned retirement.

Berlanga's contempt for the islands was such that he did not even bother to name them, nor did he make a sovereign claim for their territory. He simply referred to them as being 'worthless'. A letter to the king makes it clear that as far as Berlanga was concerned these were useless lumps of rock on which there was no water or gold, or even soil. Berlanga's none-too-subtle advice to the king was that it was best to leave them alone.

Berlanga could not even be complimentary about the islands' wildlife. He spoke to the king of having seen birds 'like those of Spain, but so silly that they do not know how to flee, and many were caught in the hand' and of tortoises so big 'that each could carry a man on top of himself'.

In making this last comment, Tomás de Berlanga became the first person ever to provide the world with a description of a hitherto undiscovered race of giant tortoises. It would, however, be many years before another human being laid eyes on either the islands or their tortoises.

The Spanish king complied with Berlanga's request and the 'worthless' islands were deliberately ignored by the conquistadors.

Only three decades later did another ship visit the islands, again after being caught in a swift-flowing current. This time the captain gave the islands a name: the Galápagos, after the crew encountered the largest 'land turtles' that they had ever seen in their lives, some of whose shells were saddle-shaped. *Galápago* is Spanish for a horse's saddle.

Here Be Giants

BY 1600 BOTH the Galápagos and the Mascarene islands featured on most nautical maps, alerting ships' captains to their presence. What would happen now that the world knew of their existence?

The Galápagos Islands had the fortune to be in such an inconvenient position that even after their appearance on a map in 1570 they received very few visitors. Some sailors were convinced that they did not exist at all or that they could magically move around the ocean, earning them the nickname of 'the enchanted islands'. Individual islands in the group remained unnamed, unmapped and unwanted by any nation. The population of 'big tortoises', 'ugly lizards' and 'silly birds' that Berlanga had seen continued to be left in peace.

Ships that did call at the islands found them to be utterly barren and waterless, and, like Berlanga, crews often departed following harrowing experiences of drought, ocean currents and weather conditions. The Galápagos were far from being an island paradise and there was no earthly reason for sailors to want to visit them. Hence, for the 200 years following their discovery they remained unexplored.

In fact, the first regular visitors received by the Galápagos were two English pirates, Ambrose Cowley and William Dampier, who took advantage of the islands' inhospitable nature and used them as a hideaway during the 1680s. During several visits to the Galápagos, Cowley and Dampier made maps of the islands and were the first to learn the number of islands in the archipelago (thirteen large ones and six smaller islets). They even named some of them after members of the English monarchy and aristocracy, such as James (now San Salvador), Charles (Santa María), Chatham (San Cristóbal) and Albemarle (Isabela). Some of these names were retained for over two hundred years before being replaced by their modern Spanish counterparts.

The pirate Cowley was not much of a writer, preferring instead to make charts, and so the physical description of the islands was left to Dampier. It is from the latter that we get the first full description of the flora and fauna of the Galápagos, including their most obvious resident, the giant tortoise:

> The Spaniards when they first discovered these islands, found multitudes of birds, and land turtle or tortoise, and named them the Gallipagos islands. I do believe there is no place in the world that is so plentifully stored with these animals. The birds here, are as fat and large, as any that I ever saw; they are so tame, that a man may knock down twenty in an hour's time with a club. The land-turtle are here so numerous, that five or six hundred men might subsist on them alone for several months, without any other sort of provision: They are extraordinary large and fat, and so sweet, that no pullet eats more pleasantly. One of the largest of these creatures will weigh one hundred and fifty or two hundred weight, and some of them are two feet, or two feet six inches over the callapee or belly. I never saw any but at this place, that will weigh above thirty pounds weight.

Dampier's description is functional, giving us an impression of the size, abundance and taste of the tortoises, but it provides no information about their appearance or behaviour. Neither the two pirates nor any other visiting sailors had any interest in the natural history of the Galápagos and, as we shall see in the next chapter, it was not until the early nineteenth century that we find the first reasonable descriptions of the tortoises. Until that time they were largely spared the attention of European naturalists.

While the remote location of the Galápagos preserved their wildlife, the same was not true of the islands of the Indian Ocean. They had the misfortune to be situated on a major trade route to India and the Far East, with a climate and soil type that were quite capable of supporting a thriving agricultural industry. The strategic position of Mauritius, Réunion and Rodriguez also meant that they were useful as military bases from which it was possible to launch seaborne attacks on the commercial shipping of a foreign enemy. Because of this, the islands were highly desirable prizes, and battles for their control were frequent, which led to many changes in the sovereignty of the Mascarene Islands throughout their history.

Mauritius, for example, was discovered in 1511 by the Portuguese, who claimed but never colonised it. They were then ousted by the Spanish, who themselves abandoned the island to the Dutch in 1598. Next came the French, who in 1715 claimed, colonised and fortified the island but were themselves forcibly removed by the British in 1810. It was much the same story on Réunion and Rodriguez, with control alternating between nations, although by the mid-eighteenth century the French had also gained control over these islands.

Because of this constant switching of sovereign power, prior to the nineteenth century the records for the Mascarene Islands are fragmentary, so that it has proved difficult to work out what happened to their tortoise populations. In fact, the best records of the tortoises and island wildlife come not from government sources but from the logbooks of ships that visited the islands, as well as from anecdotes in the memoirs of captains and other seafarers who traded, pirated or simply sailed their way around the Indian Ocean. From these we can get an idea of which islands originally had tortoise populations and how the arrival of the first ships impacted on them.

One of the earliest and most graphic descriptions of the Mascarene giant tortoises comes from the adventurer François Leguat, who wrote an account of those on Rodriguez after becoming stranded on the island for some time in 1691.

We saw no four-footed creatures but rats, lizards and land-turtles, of which there are different sorts. I have seen one that weighed one hundred pounds [forty-five kilograms], and had Flesh enough about it, to feed a good number of men. This flesh is very wholesome, and tastes something like mutton. The fat is extremely white, and never congeals or rises in your stomach, eat as much as you will of it. We all unanimously agreed, 'twas better than the best butter in Europe. To anoint one's self with this oil, is an excellent remedy for surfeits, colds, cramps, and several other distempers. The liver of this animal is extraordinarily delicate, 'tis so delicious that one may say of it, it always carries its own sauce with it, dress it how you will. The bones of the turtles are massy, I mean they have no marrow in them. Every one knows, that these animals in general are hatch'd of eggs. The land-turtles lay theirs in the sand, and cover them, that they may be hatch'd: the scale of it, or rather the shell, is soft, and the substance within good to eat. There are such plenty of land-turtles in this isle, that sometimes you see two or three thousand of them in a flock; so that one may go above a hundred paces on their backs; or, to speak more properly on their carapaces, without setting foot to the ground. They meet together in the evening in shady places, and lie so close, that one would think those places were pav'd with them. There's one thing very odd among them; they always place sentinels at some distance from their troop, at the four corners of their camp, to which the sentinels turn their backs, and look with the eyes, as if they were on the watch. This we have always observed of them; and this mystery seems the more difficult to be comprehended, for that these creatures are incapable to defend themselves, or to fly.

Monsieur Leguat provides us with by far the most detailed historical description of Mascarene wildlife and, like Dampier in the Galápagos, he is keen to point out the good taste of the tortoises' meat and oil.

In the following decades many other passing sailors would echo Leguat's description and would invariably make one of two comments about the tortoises: (1) that they could reach more than a hundred kilograms in weight; and (2) that they tasted absolutely delicious. Indeed, as time went on the passing descriptions of the Mascarene tortoises sounded more and more like a restaurant review.

Sieur Dubois wrote that the liver of the Réunion tortoise was 'one of the most delicate morsels which man can eat' and that the oil 'is marvellous for rubbing on afflicted limbs'. John Jourdain compared their flesh to the finest beef but complained that the tortoises looked so ugly that some of his men refused to eat them. In fact, there seems to have been only one person who had a bad experience dining on tortoises and that was the English aristocrat Thomas Herbert, who declared that they were 'better meat for hogs' than men. Perhaps his aristocratic palate was more refined than that of the average deckhand.

Herbert's dislike of cooked tortoise was definitely the exception and most people seem to have taken an instant liking to their meat. It was this edible quality that was to be the start of the tortoises' downfall in the Indian Ocean. As ships passed the Mascarene Islands,

In the sixteenth century the arrival of European sailors on Mauritius proved to be bad news for its native animals, especially the tortoises.

they would stop off and feast on the tortoises, taking some on board as live provision for the remainder of their journey. Their chief reason for doing so, other than the wonderful taste of the meat, was that the tortoises could be stored live on deck for long periods without needing to be either fed or watered. In an age when fresh food could not be easily stored at sea, the tortoises provided a portable supply of fresh meat. To sailors who had been otherwise surviving on salted meat, pickled vegetables and biscuits, the tortoises were a godsend.

In the early years an occasional ship calling in and taking a few tortoises had little impact, but as the trade routes opened up, so the number of visiting ships increased dramatically, putting pressure on the population.

It is possible to work out roughly what was happening to giant tortoises in the Indian Ocean from entries made in ships' logbooks. From these we can be absolutely certain that during the early seventeenth century there was an abundance of tortoises on Mauritius, Réunion and Rodriguez. In 1630 Herbert described seeing on Mauritius 'land tortyses, so great that they will creepe with two mens burthen', while in 1630 Henri du Quesne noted that Réunion had vast numbers of giant tortoises so big that they could 'carry a man with more ease than a man can carry them'.

These reports can leave us in little doubt that all three islands in the Mascarene group had native giant tortoises but as seafarers ventured further, native populations were found in two other localities in the Indian Ocean. In 1609 the Seychelles archipelago was discovered and was found to have its own native giant tortoise, but the task of unravelling the animal's natural history has proved very difficult.

There are over 115 islands in the jurisdiction of the Seychelles, most of which are small coral atolls only a few metres above sea level and sparsely vegetated, although the main islands are composed of resistant granite and therefore quite mountainous. Initially it was thought that native giant tortoises were found only on the larger granite islands but visiting ships during the seventeenth century seem to mention their presence on practically every Seychelles island. However, this is thought to be unlikely because most of the atolls were too dry and barren to support large animals.

Faced with this confusion, it has taken modern naturalists some time to distinguish what are likely to be mistaken reports of Seychelles tortoises from true ones. The most recent research makes it seem probable that when the Seychelles were first discovered only three of the islands had populations of tortoises. These were Mahé and North Island, which are both granitic, and Astove Island, which is an atoll.

There have been reports at various times of living or fossil tortoise material on another eleven Seychelles atolls but it is thought that these are probably either erroneous, the remains of animals that are long extinct or simply individual tortoises that have been dumped there by passing ships.

After the discovery of tortoises in the Seychelles, the final island inhabited by these giants did not come to light until 1744. This was the low coralline atoll of Aldabra, which lies in a little-visited part of the Indian Ocean about 400 kilometres north of Madagascar. Aldabra was discovered to have a race of indigenous giant tortoise but its small size and remoteness meant that it received only two traceable visitors during the whole of the eighteenth century. The last of these noted that there were a 'great many land-turtle much larger than those at Rodrigues'.

Thanks to ships' logbooks, we know that, of the many hundreds of islands in the Indian Ocean, it can now be said for certain that the three Mascarene islands, Aldabra and at least three islands in the Seychelles had wild populations of giant tortoise at the time of their discovery. Of course, the Galápagos also had their own native giant tortoises, which then spread across an unknown number of islands.

Discovering these animals was one thing. Understanding how they wound up living in the middle of nowhere was something else again. The occurrence of gigantic tortoises in two so completely separate geographical locations presented the natural history community with the inevitable question: were the giant tortoises on the Galápagos and Indian Ocean islands related to one another or were they two different species?

In either case the answer raised more issues than it settled. If they are one species, how did they come to be so far apart from one another? If they are separate species, how did two identical-looking animals end up in similar kinds of locations on opposite sides of the globe?

In attempting to answer this question one person would stumble upon one of the most fundamental and all-encompassing discoveries in scientific history.

PART TWO

Inspiration

Just One Species?

I N 1735 THE Swedish botanist Carl Linnaeus published a pamphlet entitled *Systema Naturae* in which he proposed a universal means by which plants and other organisms could be classified. Until then the scientific naming of plants and animals had been a random affair with no set convention about what exactly constituted a new species or, indeed, how one should be named. Many plants and animals had been classified several times over by different scientists; the names themselves were often long, rambling Latin sentences. This, combined with poor anatomical description, meant that it was very difficult for scientists to distinguish plant and animal species that had been named from those that had not. The result was a chaotic mess that obscured the evolutionary relationship between different species around the globe.

Unlike most of his contemporaries, when Linnaeus contemplated the spectrum of life on earth he was certain that in it he could see a hidden hierarchical order. Wherever he looked he could see similarities between many types of plant and animal, and he believed that they could be classified in small groups. He also believed that these groups could themselves be subdivisions of larger groups. For example, lions, leopards and tigers are obviously different species, yet they look similar enough to one another to warrant their being collectively called 'big cats'. Many of the features of the big cats (for example, elongated body shape, sharp teeth, sharp claws, fur) are shared by other animals such as dogs, ferrets and bears, suggesting that there is a loose relationship between them. (Nowadays we recognise that cats, dogs, ferrets and bears belong together in the Order Carnivora.)

Evidence like this, thought Linnaeus, was proof of God's hand at work, and he hoped that by creating a classification system that mimicked God's blueprint for nature, mankind could better understand the mind of the Creator. In *Systema Naturae* Linnaeus proposed that each distinctive plant or animal be designated a place in a classification scheme which could reflect its relationship with the rest of nature.

For example, within Linnaeus' system humans have the species name *sapiens* within the genus *Homo*, which is itself within the Family Hominidae (gorillas, chimpanzees and humans), which is within the Order Primates (all the monkeys and apes), which is itself within the Class Mammalia (all mammals), which itself lies within the Kingdom Animalia (the whole animal kingdom). Thus, using Linnaeus' system, we can see *Homo sapiens'* relationship to the whole of the rest of nature.

Systema Naturae was to go through many changes before it was adopted as the standard means of classifying living and fossil organisms but by the time of Linnaeus' death in 1778 his was the universal means by which naturalists could name and describe their new discoveries. The Linnaean system is still with us today and is backed up by a formidable set of rules and regulations, which are themselves enforced by a powerful international committee of natural scientists who stand in judgement on disputes, loopholes or revisions thrown up by the system.

One result of the acceptance of the Linnaean classification system was an immediate rush by many amateur and professional scientists to name and describe any and every plant and animal according to the new rules. This coincided with a period of unprecedented overseas exploration, expansion and colonisation by the European nations, whose ships were now finding their way into every nook and cranny around the world. Sometimes young scientists would join these voyages, returning with armfuls of strange animals and plants that they had collected on the way. In a bid to further their career, the returning scientists would often donate their carefully gathered specimens to museums. Here they would be picked over by the resident curators, many of whom were eager to be the first to put a name to these new creatures.

As increasing numbers of scientists fanned out across the world, it could only be a matter of time before giant tortoise specimens would be collected, studied and given an official name under the Linnaean system.

Many scientists were already well aware of the existence of the giant tortoises from the short anecdotal accounts in the journals of ships' captains. Even so, none of the descriptions had been clear enough to permit scientific classification. In fact, as we have already seen, the majority of accounts of the tortoises describe the taste of their meat rather than any physical characteristics. Besides, under the Linnaean system it was necessary to nominate a single physical specimen as the 'holotype' to which all other identifications would have to be compared, and as yet no giant tortoise specimens had made their way into the correct hands. For these reasons the giant tortoise was not to receive a scientific name until almost exactly three hundred years after the first human sighted them.

In the year 1812 the German botanist August Friedrich Schweigger was given a large, smooth shell that had once belonged to a giant tortoise. Exactly where this giant specimen had originated was unknown to Schweigger although he suspected that it hailed either from the Indies or possibly Brazil.

Schweigger was more used to dealing with plants than animals but he was quick to recognise that this was the carapace of a tortoise far larger than any hitherto known, and that no such animal had been described by anybody else. Naming new species was a matter of pride among the scientific community, and so even though reptiles were not his speciality, Schweigger was not going to miss the opportunity to define such a magnificent animal. Hence the specimen was afforded the name *Testudo gigantea*, which means 'gigantic tortoise' in Latin.

In the absence of any information to the contrary, Schweigger assumed that the many sailors' descriptions of giant tortoises in the Indian Ocean referred to his new species. Thus the tortoises of the Seychelles, Aldabra, Mauritius, Réunion and Rodriguez were all now classified under the name *Testudo gigantea*.

The same logic was applied when news of the Galápagos giant tortoises started to filter back to Europe and America. With no actual specimens to determine otherwise, it was simply assumed that the tortoises of the Indian Ocean and the Galápagos Islands were the same species: *Testudo gigantea*.

In order to explain how two populations of giant tortoise could exist so far apart, it was widely accepted that the animals from the Indian Ocean had been transported to the Galápagos by passing sailors. It was, after all, common practice for ships to pick up tortoises in Mauritius and keep them on board for weeks or months as a food supply. One or more ships must have carried Mauritius tortoises on board whilst they traversed the Pacific Ocean, later abandoning them on the Galápagos Islands.

Thus by the 1820s those few scientists who were interested in the giant tortoises believed that there was just one species that had, through the actions of man, become dispersed across two oceans. Schweigger's name, *Testudo gigantea*, was thus liberally applied to any new giant tortoise specimen arriving back in Europe, regardless of its provenance.

∿

Strangely, while the European naturalists were in agreement with one another about the existence of only one giant tortoise species, many of the sailors and captains who had seen these animals first-hand were not so convinced and thought that they could see subtle differences between the various populations. Had the European naturalists read some of the eyewitness accounts from visitors to the Galápagos and Indian Ocean islands, perhaps they would have thought differently about the all-encompassing status of *Testudo gigantea*.

The first hint that there might be more than one species of tortoise comes from the pen of the American merchant trader, Amasa Delano. He circumnavigated the globe several times during his life, visiting the Galápagos and Mascarene Islands on many occasions.

Delano's memoirs, which were published in 1817, contain an entertaining account of his voyages but they also hold one of the first really detailed anatomical descriptions of the Galápagos tortoises and their behaviour. Delano's description of his first encounter with a giant tortoise is typical of his acute eye:

> I was put in the same kind of fear that is felt at the sight or near approach of a snake, at the first giant tortoise I saw, which was very large. I was alone at the time and he stretched himself as high as he could, opened his mouth, and advanced towards me. His body was

raised more than a foot from the ground, his head turned forward in the manner of a snake in the act of biting, and raised two feet and a half above its body. I had a musket in my hand at the time, and when he advanced near enough to reach him with it, at the touch of which, he dropt himself upon the ground and instantly secured all his limbs within his shell.

Aside from this encounter, it was Delano's remarks about the worldwide occurrence of the giant tortoise that were most important: 'I have seen them at one or two other places only. One instance was those brought from Madagascar in the Isle of France [Mauritius]; but they were far inferior in size, had longer legs, and were much more ugly in their looks than those of the Galápagos Islands. I think I have likewise seen them at some of the Oriental Islands which I visited.' Although he was no biologist, Delano was convinced that the tortoise populations of the Indian Ocean and Galápagos were physically different from one another.

Similarly, a couple of years prior to Delano's account, the memoirs of a captain in the American Navy, one David Porter, were turned into a book. Porter had visited the Galápagos Islands many times and had consequently developed a fascination with the wildlife there, especially the tortoises. Among his many descriptions of them, Porter appears to have noted what he thought were differences between the populations on different islands:

The shells of those of James [San Salvador] Island are sometimes remarkably thin, and easily broken, but more particularly so as they become advanced in age; when, whether owing to the injuries they receive from their repeated falls in ascending and descending the mountain, or from injuries received otherwise, or from the course of nature, their shells become very rough, and peel off in large scales, which renders them very thin and easily broken. Those of James Island appear to be a species entirely distinct from those of Hood [Española] and Charles [Santa María] islands. The form of the shell of the latter is elongated, turning up forward in the manner of a Spanish saddle, of a brown colour and of considerable thickness. They are very disagreeable to the sight, but far superior to those of James Island in point of fatness, and their livers are considered the greatest delicacy. Those of James Island are round, plump, and black as ebony, some of

Captain David Porter's 1822 sketch of a Galápagos tortoise. He was one of the first people to suggest that there may be discernible differences between the world's giant tortoises.

them handsome to the eye, but their liver is black, hard when cooked, and the flesh altogether not so highly esteemed as the others. [The Española Island tortoises] were of a quality far superior to those found on James Island. They were smaller in appearance to those of Charles Island, very fat and delicious.

Taken together, the observations of Delano and Porter would appear to indicate not only that the Indian Ocean and Galápagos giant tortoises could be separate species, but also that there might well be more than one species living within the Galápagos archipelago.

Was Schweigger's *Testudo gigantea* a single giant tortoise species that covered the entire globe or just one of several species worldwide? The truth is that in the early 1800s few people lost much sleep over this question.

The islands of the Indian Ocean and the Galápagos were thousands of kilometres from Europe, and although some specimens were finding their way back to museums and universities, nobody had a sufficient number of them to make a meaningful comparison. Even if they had, few of the specimens had been labelled, making it impossible to tell which island (or in some cases which ocean) they had come from. For the time being, the academic world was happy to go along with Schweigger's all-encompassing nomenclature of *Testudo gigantea*.

In a period of discovery and new ideas, the problem of giant tortoise speciation was small beer in comparison with other heated debates taking place elsewhere in the scientific community. One particularly virulent row had occurred over a new French-inspired theory that was nicknamed 'evolution', the main implication of which was that plant and animal species were capable of spontaneously changing aspects of their shape and biology through time, eventually creating new species. Evolution was considered a heresy and only a few people believed in the idea. In fact, most scientists were doing everything in their power to stamp it out.

In the years to come, the humble and forgotten giant tortoises of the world were to play a pivotal role in the understanding of what evolution is and how it operates. It was all thanks to a shy young man named Darwin, who had a close encounter with a giant tortoise that inspired him towards greater things.

A Tortoise for Each Island

IN THE LATE autumn of 1830, Captain Robert FitzRoy of His Britannic Majesty's surveying ship *Beagle* returned to England after spending several years surveying the South American coastline. The voyage was considered to have been a great success and within months Captain FitzRoy was campaigning to be allowed to take the *Beagle* back to the South Atlantic to finish its survey of the Patagonian coastline. After he had used his family connections (FitzRoy was the descendant of a duke) to put pressure on the Admiralty Board, the new expedition was given the go-ahead and FitzRoy set about assembling a crew for departure in October 1831.

During the previous voyage Captain FitzRoy had become fascinated with the living world around him and was frustrated that there was not a scientist on board who could observe and interpret the great wonders he had seen. FitzRoy therefore created the post of ship's naturalist so that he might have at hand a learned companion with whom he could share his own curiosity. However, only weeks before departure the position of naturalist remained vacant, having been initially accepted, but later refused, by two eminently suitable candidates, one because his wife 'looked so miserable at the prospect'.

As he was about to despair FitzRoy heard, through a friend, of a talented young naturalist from a well-respected background who might just suit the post. The young man's name was Charles Darwin.

When at the age of twenty-two Darwin received the invitation to travel on the *Beagle*, he was preparing himself for a career in the clergy. He had already abandoned his medical training from fear of the sight of blood but his increasingly agnostic thoughts showed that the Church was not exactly ideal either. His first love was the natural world and he was already an accomplished, self-taught zoologist, but even so his father did not approve of having a naturalist son. At first, therefore, Charles felt obliged to turn down FitzRoy's offer.

Fortunately Darwin Senior relented and in late October 1831, Charles, FitzRoy and the rest of the *Beagle*'s crew all converged on Plymouth, although bad weather meant that the ship did not actually leave until after Christmas.

FitzRoy was glad to have Darwin on board for company but there had actually been another motive behind his desperate desire for the *Beagle* to have a ship's naturalist. In addition to being fascinated by science and nature, FitzRoy was also a man of strong religious conviction. He had been profoundly disturbed by recent talk about the supposed ability of plants and animals to 'evolve' from one species to another. If true, then 'evolution' would fly in the face of the biblical account of the Creation where all life on earth was created by God in only a few days. To suggest that a species might be able to change spontaneously into another, entirely different species would be to cast doubt on the whole idea of the Garden of Eden. FitzRoy had purposely sought out a ship's naturalist in the hope that he might be able to find evidence wholly in favour of the biblical idea of Creation. Darwin, a young trainee priest with sound scientific credentials, seemed to be just the sort of fellow for the task. With hindsight it is difficult to see how FitzRoy could possibly have picked a worse candidate for his clandestine mission against evolutionary science.

⁓

The idea that a species could evolve was not new. The French had pursued it keenly during the 1790s and even Charles Darwin's grandfather Erasmus had tinkered with the notion, but 'evolution' did not gain much scientific credibility and was dismissed by all but a few zealots. Then, in the early 1830s, the issue of evolution was revived after the geologist Charles Lyell published the first volume of his *Principles of Geology*, a work that is still held dear by many modern earth scientists.

In *Principles*, Lyell sought to demonstrate that there were links between all the observable processes operating on the earth today (such as volcanoes, earthquakes, tides, weather, etc.) and some of the features that geologists found in ancient rocks. Lyell opined that it was possible to look at the physical characteristics of rocks and, by comparing them with the processes operating on the modern earth, deduce what sort of environmental processes had produced them.

'Although we are mere sojourners on the surface of the planet,' wrote Lyell in his first volume, 'chained to a mere point in space, enduring but for a moment of time, the human mind is not only enabled to number worlds beyond the unassisted ken of mortal eye, but to trace the events of indefinite ages before the creation of our race, and is not even withheld from penetrating into the dark secrets of the ocean, the interior of the solid globe; free like [a] spirit.'

To Creationists like FitzRoy, Lyell proposed a whole collection of unpalatable ideas, such as his belief that the world must be far older than the few thousand years often quoted by biblical scholars. Crucially, Lyell also held that 'there is a progressive development of life, from the simplest to the most complicated forms' and that 'man is of comparatively recent origin'. While Lyell at no point mentioned evolution, the evidence that he was offering certainly pointed in that direction.

It was to counter exactly this kind of heretical talk that FitzRoy had sought out a person such as Darwin. What a bitter disappointment it must have been for the captain to have watched the young naturalist walk on board the *Beagle* with a copy of Lyell's *Principles* tucked under his arm.

Contrary to popular belief Darwin did not embark on the voyage intent on revolutionising the scientific world. He had been brought up and trained along lines that were not far removed from those of his Creationist captain. He did, however, have an active mind and, thanks to his upbringing, an inquisitive disposition. Hundreds of ideas would come to him each day, few of which he ever completely dismissed and certainly not on the grounds that they were inconvenient to his religious, scientific or political beliefs.

Nevertheless, during his five years aboard the *Beagle* Darwin would have several experiences that would each eventually contribute to the realisation that one species could spontaneously evolve into a new and completely different species.

The first of these was his reading of Lyell's *Principles of Geology*, in between bouts of chronic seasickness. As the *Beagle* docked at various volcanic islands, Darwin saw with his own eyes that Lyell's idea that sea levels had been higher in the past was probably true. How else could Darwin explain the perfectly preserved fossils of seashells stuck halfway up the side of a volcano? The timescales necessary to accommodate such changes awakened in Darwin the idea that the earth was much older than he had been led to believe.

Next came his visit to inland Argentina where he discovered fossils of bizarre-looking giant armadillos and ground sloths. These animals were obviously no longer represented on the earth, confirming for Darwin the concept that species could become extinct. Initially, he went along with the idea that this extinction could be put down to biblical cataclysms such as Noah's Flood. It was not until the arrival of volume two of *Principles of Geology*, posted to Darwin from England by a friend, that he began to ditch his old biblical beliefs.

The second volume of *Principles* was entirely devoted to the relationship between animals and plant species and their landscape. Lyell argued that each animal was adapted perfectly to live in a particular environment and that changes in this environment might lead to extinction. He painted a picture in which through the expanse of geological time the earth's landscapes had undergone continual change, and as some environments disappeared, those plants and animals that lived in them would vanish too. As new environments emerged, so new species would populate them. Thus, looking back through geological time, there was a continual process of old species becoming extinct with the vacant slots they left being filled by newer species.

While this might sound evolutionary, Lyell was adamant that 'a transmutation of species' was absolutely not possible. Just how a new species could be created, Lyell could not say but even the strong evidence he put forward afforded 'no ground for questioning the instability of species'.

At first Darwin had agreed with his mentor's views, but as the voyage progressed and he encountered 'varieties' of plant and animal, he was not so sure. In Darwin's time a 'variety' was an animal or plant that, while being recognisable as a particular species, had a characteristic that made it distinctly different from other members of that species. That characteristic could be anything: height, colour, shape, etc. Most species included several distinguishable varieties, with perhaps the best examples being in botany where the same species could encompass varieties that produce flowers of different colours (for example, the many varieties of rose so prized by gardeners).

Darwin looked at the many 'varieties' that he was seeing and speculated that the physical characteristics of an individual species might not be fixed in stone as Lyell had suggested. Was it not possible, thought Darwin, that a species could vary so much that some of the more extreme 'varieties' budded off into entirely new species? Where does one draw the line between a variety within a species and an entirely new species?

Darwin began to depart from Lyell's set text and started to think about the fluidity of species. He had taken his first step down the rocky path that led towards evolutionary theory but he still had a long way to travel.

In the spring of 1834 the *Beagle* rounded Cape Horn and began ploughing her way north up the Chilean coast. In the preceding three years, Darwin's whole philosophy regarding the natural world had been turned on its head.

Gone were the old ideas of a young earth of divine creation populated by fixed species. In its place was a vision of a geologically active and turbulent planet whose plants and animals could become extinct, with new ones appearing in their place. Lyell's books and his own experience had brought to him these conclusions but there was still a great void in his understanding. If old species could become extinct and new ones take their place, where on earth did these new species come from?

Darwin's experiences aboard the *Beagle* had answered so many questions already but would they really find a reply to this most fundamental of questions?

~

On 12 February 1832, while the *Beagle* was making the treacherous crossing between West Africa and Brazil, the Galápagos Islands changed sovereignty and, from being nominally British, became

The young Charles Darwin, whose encounters with Galápagos tortoises
were to start him thinking about the nature of evolution.

firmly Ecuadorian. Two years later, as the *Beagle* slowly wound her way around South America, surveying the nooks and crannies of its crenulated coastline, Ecuadorian settlers arrived at the Galápagos, most of them exiled criminals. In the absence of any volunteer settlers, the Galápagos Islands had become a penal colony.

On 7 September 1835 the *Beagle* set off from Peru on a northerly course. After nearly four years, her survey of South America had been completed and the ship could now begin the long voyage home via the Pacific Ocean and its many islands.

However, the *Beagle's* crew had one further duty to perform on their way back to England: to stop at the Galápagos en route in order to survey some of the larger islands. The visit would also be a good excuse to stock up on food and water before undertaking the four-week journey to the South Pacific.

After nine days' sailing, the *Beagle* arrived in the Galápagos where she was to undertake surveys of San Cristóbal and Española islands before continuing to Santa María.

Four years of travel on board the *Beagle* had revealed a wealth of natural history marvels to Darwin. His naturally inquisitive mind ensured that his notebooks were by now full of valuable descriptions, while his compatriots in London and Cambridge had already received boxes full of plants, animals and fossils, many of which were new to science. He was still at a loss to understand just how a species could be spontaneously created but in the autumn of 1835 this problem was not foremost in his mind.

Despite his successes, Darwin's spirits were then at an all-time low. In the previous months he had to endure extended stays in a number of lawless Peruvian towns, ending with one in the anarchic city of Lima, which he described as being in 'a wretched state of decay'. He was tired of the heat, filth, noise and thuggery that were synonymous with the South American frontier and longed to be back in his family's country house in England. Before departing from Lima he had written to his cousin complaining that 'This voyage is terribly long. I do so earnestly desire to return, yet I dare hardly look forward to the future, for I do not know what will become of me.' The constant seasickness had given way to homesickness, coupled with anxiety about the uncertain job prospects on his return.

As the *Beagle* approached the Galápagos Islands it was not thoughts of animals that preoccupied Darwin but stories of fiery volcanoes and thick lava fields that he had heard from sailors who had previously visited these islands. He was not to be disenchanted, but after his first full day there he seems to have been surprised by the sheer starkness of the landscape. In his diary he records that 'The black rocks heated by the rays of the vertical sun like a stove, give to the air a close and sultry feeling. The plants also smell unpleasantly. The country was compared to what we might imagine the cultivated parts of the Infernal regions to be.' His hope was that the Galápagos volcanoes, being geologically immature, might have preserved on their slopes some fossils from recently extinct animals. Previous visits to volcanic islands had produced fossilised seashells from high up these slopes, helping to convince Darwin that Lyell's hypothesis regarding changing sea levels was correct.

He was to be sorely disappointed. The harsh climate and lava landscape meant that the two essential prerequisites for fossil preservation, water and sediment, were almost entirely absent. It was a fossil-hunter's worst nightmare come true.

The fossils may have been in short supply, but the Galápagos was home to a fair amount of wildlife, little of which had ever been formally described by a scientist. As the *Beagle* travelled around and between the islands, Darwin would go ashore with his servant and set about collecting plants, shells, birds and reptiles while making notes in both his daily diary and his zoological notebook

～

It was on 24 September that the *Beagle* anchored in the harbour on Santa María Island, with the intention of resting there for a few days to replenish the ship's supplies and also to give the crew a break ashore. By chance the Galápagos' acting governor, Captain Nicholas Lawson, had wandered down to the harbour to visit a whaling ship that was also anchored there.

Lawson, an Englishman working for the Ecuadorian government, was doubtless delighted to have chanced upon the *Beagle*, a ship full of his fellow countrymen, some of whom were actually men of breeding (unlike the rough-and-ready crew on the whaling boat). At last the chance of a decent conversation!

The next morning Lawson returned to the *Beagle*, volunteering to take anybody who was interested into the mountains to visit the 'settlement' (or, more correctly, penal colony). Darwin, FitzRoy and several others accepted the offer. That trek would change Darwin's life for ever.

The dramatised introduction to this book describes the journey made by Darwin and his fellow shipmates up to the Santa María Island settlement on 25 September. The party of men was evidently well treated by Governor Lawson and the day's events receive a much longer write-up than usual in Darwin's diary. Everything about the visit, from the geology to the settlers themselves, seems to have fascinated him and remained in his memory long enough to be transcribed. In particular, it would appear that at some point during the day Darwin had a one-to-one conversation with Lawson and that the two men talked at length about two of Darwin's favourite topics: food and natural history.

On the Galápagos the subject of food is synonymous with the subject of tortoises, so Darwin and Lawson ended up talking tortoise for some time. It was an area on which the governor was very knowledgeable, as Darwin's scribbled notes reveal:

> The main article of animal food is the terrapin or tortoise: such numbers yet remain that it is calculated two days hunting will find food for the other five in the week. Of course the numbers have been much reduced; not many years since the Ship's company of a Frigate brought down to the Beach in one day more than 200 – where the settlement now is, around the Springs, they formerly swarmed. Mr Lawson thinks there is yet left sufficient for twenty years: he has however sent a party to James [San Salvador] Island to salt (there is a Salt mine there) the meat. Some of the animals are there so very large, that upwards of 2 hundredweight [about 100 kilograms] of meat have been procured from one. Mr Lawson recollects having seen a Terrapin which six men could scarcely lift & two could not turn over on its back. These immense creatures must be very old: in the year 1830 one was caught (which required six men to lift it into the boat) which had various dates carved on its shell; one was 1786. The only reason why it was not at that time carried away must have been, that it was too big for two men to manage. The Whalers always send away their men in pairs to hunt.

Darwin was impressed with Lawson's knowledge of the tortoises and eagerly wrote down every titbit about them that was thrown to him. It was during their wide-ranging conversation that Lawson commented that he could tell which island a particular tortoise had come from, simply by looking at the shape of its shell. Darwin thought nothing of this comment at the time and omitted it from his diary, although he did later jot it down in his zoological notebook.

It would take months for him to realise the remark's significance but because of it future generations of biologists would have good cause to thank providence for bringing together on that day the young naturalist and the boastful governor. In time Governor Lawson's remark would be the spark that started a fully fledged scientific revolution.

~

It is quite possible that natural history discussions with Governor Lawson prodded Darwin out of his homesickness and into naturalist mode once more. During the rest of his time in the Galápagos, he and his servant diligently collected plants and animals from every island that they visited.

The tortoises continued to attract Darwin's attention and his notes about them lengthened. Although the existence of the giant tortoises had been known since the sixteenth century, Darwin was the first person to make a serious study of their behaviour. From his notes it was for the first time possible to gain an idea of what everyday life for a giant tortoise on the Galápagos was like. Indeed, so detailed are his observations that he must have quizzed not just Governor Lawson but a good many other people about the tortoises' habits, as he could not possibly have had the time or good fortune to observe them eating, mating, dying, egg-laying and all the other events in their lives that he describes.

Darwin's fascination began in fact from the moment of his first landing on the Galápagos Islands. On disembarking he was puzzled to observe that there were dozens of wide and well-trodden tracks that criss-crossed the island, forming what looked like a primitive road network.

'When I landed at Chatham Island [San Cristóbal],' he wrote, 'I could not imagine what animal travelled so methodically along the well-chosen tracks.' But after following one of the paths for a while he soon discovered that they had been created by the giant tortoises, which would use them to navigate to and from watering holes, and also to get from the highland areas to the lower coastal regions.

On following one tortoise track to a watering hole, Darwin watched as several of the huge beasts jostled for space around the edge of a small muddy puddle. Periodically one would plunge its head entirely into the dirty water and then take several slow, deep gulps until its thirst was satisfied. The young naturalist observed that one of the secrets of the tortoises' success was their ability to store water in a special bladder, on which they could survive for months. Furthermore, in an emergency, this water could be drunk by thirsty sailors. Having tried it himself, Darwin declared the water to be 'only slightly bitter'. For food, he observed, the tortoises would eat cacti or the leaves and berries of certain trees.

Darwin found that the tortoises were quite deaf and that, as long as they could not see him, they would continue about their business, allowing him to observe and even conduct experiments with them. Darwin was fascinated by their ability to cover such long distances and began matching them pace for pace over measured lengths. By doing so he calculated that they could cover perhaps seven kilometres in a day – quite a feat for an animal weighing several hundred kilograms.

Even the tortoises' mating habits did not go unnoticed. 'The males copulate with the females in the manner of a frog [and] remain joined for some hours. During this time the male utters a hoarse roar or bellowing, which can be heard at more than 100 yards [91 metres] distance. When this is heard in the woods, they know certainly that the animals are copulating.' The eggs would be laid months later, buried in shallow pits or, as Darwin observed on one occasion, 'in a kind of crack'. The hatchlings were often prey to birds but adult tortoises had only ever been observed to die by accident and not through natural causes.

Darwin's tortoise observations have been proved to be remarkably accurate and, whilst similar observations were made by later expeditions, it was not until a research station was founded on the Galápagos in the 1960s that detailed scientific notes would be made on their biology and behaviour.

It is important to note that the tortoises of the Mascarene Islands and the Seychelles became extinct in the wild before any detailed observations of their behaviour could be made. The first observations of the Aldabra tortoise in the wild were not noted until the 1970s; they proved to have remarkably similar lifestyles and habits to those found on the Galápagos.

∼

Darwin may have made copious notes on the tortoises but he did not trouble to collect any from the four islands that he visited. Apart from anything else, he had no need to as the crew of the *Beagle* had twice been sent ashore to San Cristóbal Island to gather live tortoises as food for the long Pacific journey ahead. By the time the *Beagle* set off from the Galápagos, on 20 October, her deck held over thirty adult tortoises whose only purpose was to provide a fresh source of meat for Darwin and his companions as they made the four-week crossing to Polynesia.

By good fortune not all the tortoises on board were destined to end up inside the sailors' stomachs. In among the clutter of Darwin's cramped cabin lay the only pet that he was ever to adopt during the five-year voyage – a small tortoise that, as an afterthought, he had taken from San Salvador Island.

Captain FitzRoy, ever inquisitive about the wildlife around him, had also taken two small tortoises from Española Island. Unlike Darwin, who never once refers to his pet tortoise in his diary or zoological notes, FitzRoy became obsessed with his small charges, patiently measuring their growth rate and keeping notes on their appearance in his private journal. Yet another pet tortoise, from Santa María Island, was held by Darwin's servant, Syms Covington.

While the crew steadily ate their way through the adult tortoises on board, tipping their shells and boiled bones overboard, the four pets remained safe and sound below decks for the remainder of the voyage.

As he left the Galápagos behind, Darwin expressed his displeasure at the islands. 'I should think,' he wrote in his diary, 'it would be diffi-cult to find in the intertropical latitudes a piece of land 75 miles [120 kilometres] long, so entirely useless to man or the larger animals.'

As the *Beagle* sailed westwards across the Pacific, Darwin's thoughts kept returning to the time he had spent with Governor Lawson. There was something about his day at the Santa María Island settlement that kept nagging at the back of his brain. By the time he realised what it was, it was too late. He had made several crucial errors and the only way to rectify them would require returning to the Galápagos, something that he would never get the chance to do.

On 2 October 1836 the *Beagle* dropped anchor in the English port of Falmouth, ending her prolonged voyage around the world. The journey that FitzRoy had initially estimated as lasting two years had in fact taken almost five. Even the captain, the hardened sea dog more at home on a ship than on land, was pleased to have completed his tour of duty. Both he and Darwin had had an extended period to reflect on what to do with their lives once back in England and both were now ready to implement the plans they had formulated.

In the case of FitzRoy, this meant getting married (much to his shipmates' surprise) and expressing his hard-line religious leanings by lay preaching. Darwin, on the other hand, decided to forsake the uncertainty of his former life and go in all-out pursuit of a career as a naturalist.

The *Beagle* voyage had given him a new sense of direction but he was still very much a rebel in search of a cause. His mind was swimming with all that he had seen during the previous half-decade and his behaviour suggests that he was aware that there was much more to the natural world than conventional science would have him believe. The voyage of the *Beagle* had furnished Darwin with all the information and experiences he needed to solve some of the most outstanding questions in nature. Now it was up to him to assemble the data into a coherent order.

Revelation

IN RECENT DECADES there has been a quest to understand when, where and how Charles Darwin made the mental leap that converted him from an essentially Creationist point of view to that of a pro-evolutionist.

For the best part of a century it was commonly believed that, when he finally stepped off the *Beagle* and on to British soil, he had already been converted to the idea of evolution. Another part of this scientific folklore holds that Darwin's visit to the Galápagos Islands had such a profound and instant effect on him that it became the catalyst for some kind of evangelical experience so that, by the time he left the islands, he had formulated the basic ideas that would later be expanded into his greatest achievement, *The Origin of Species by Means of Natural Selection*.

One apocryphal story that occurs time and again is that as the *Beagle* left the shores of the Galápagos a spectacular row erupted between Darwin and FitzRoy. Despite Darwin's passive nature such rows had occurred before, the worst taking place in Brazil when the young naturalist had challenged the captain's justification of black slavery. FitzRoy had reacted with such ferocity that Darwin thought that he would be thrown off the ship and sent home. Fortunately FitzRoy's moods would disappear as suddenly as they arrived and he was always quick to make amends for any ill feelings caused.

The supposed subject of the post-Galápagos row was the finches that were abundant on the island. As ever, FitzRoy had been making his own natural history notes and confided to Darwin that he had noticed the different shapes of the finches' beaks. This, he concluded, was because there were many different species specifically 'created' (supposedly by God) for each island. Darwin disagreed, saying that there were not in fact many different species but that all the finches were varieties of one species that had become 'adapted' to the particular conditions on each island. On hearing this FitzRoy is alleged to have exploded, calling Darwin a blasphemer and yet again questioning his place on the ship.

Thanks to stories like this, it was usual to argue that Darwin's conversion to evolutionary science had been triggered by his study of the Galápagos' finches and the realisation that the different species had become perfectly adapted to life on their own particular island. His finches have gone down in history for their unwitting, but invaluable, contribution to science. Thus within a few days of leaving the Galápagos, Darwin was alleged to be well on the path already to recognising both the reality of evolution and that it was a problem that needed solving.

As an old man Darwin seemed to endorse this version of events when, in his autobiography, he wrote the following in order to explain how he became an evolutionist: 'I had been deeply impressed by the South American character of most of the productions of the Galápagos Archipelago, and more especially by the manner in which they differ slightly on each island of the group; none of these islands appearing to be very ancient in a geological sense.'

For many years it was believed not only that the Galápagos finches were directly responsible for Darwin's conversion, but also that the idea had come to him whilst he was still in the islands themselves. In fact neither notion is true, but it took researchers decades to work out what really happened to inspire Darwin to take the road of evolutionary science. It is not the finches, but in fact the giant tortoises, that we must thank.

~

Rather than arguing with FitzRoy about finches, as the *Beagle* began its slow westward path across the Pacific its resident naturalist lapsed into one of his periodic phases of inertia. In a later letter to a friend, Darwin confessed: 'I last wrote to you from Lima, since which time I have done disgracefully little in Natural History; or rather I should say since the Galápagos Islands, where I worked hard.'

It was not until nine months later, in the summer of 1836, as the *Beagle* made her homeward voyage across the Atlantic, that Darwin began the tedious process of assessing all that he had seen and collected during the previous five years. It was at this point that he first noticed an unusual trait displayed by some of his Galápagos specimens. These were not the famous finches, which, as we shall see, were in a confused jumble with incomplete labels attached, but instead in the mockingbird specimens, which had been collected and correctly labelled by his servant Syms Covington.

After re-examining the mockingbirds, Darwin recorded in his zoology journal:

> The [mockingbird] specimens from Chatham [San Cristóbal] and Albemarle [Isabela] Island appear to be the same; but the other two are different. In each island each kind is exclusively found: habits of all are indistinguishable. When I recollect, the fact that from the form of the body, shape of scales and general size, the Spaniards can at once pronounce, from which island any tortoise may have been brought. When I see these islands in sight of each other, and possessed of but a scanty stock of animals, tenanted by these birds, but slightly differing structure and filling the same place in Nature, I must suspect they are only varieties. The only fact of a similar kind of which I am aware, is the constant asserted difference – between the wolf-like fox of East and West Falkland Islands. If there is the slightest foundation for these remarks the zoology of archipelagos will be well worth examining; for such facts would undermine the stability of species.

This brief note in his zoology journal is generally agreed to be the first clue that Darwin was entertaining the pro-evolutionary concept that one species might be able to transmute into another, completely distinct species.

At a stroke, this statement makes it clear that, even though he was almost at the end of his voyage, Darwin was at that point by no means convinced of the case for evolution. In truth, however, it is difficult to know what he is actually trying to say. Although he mentions that species may be unstable, he does not make it clear whether he is supporting or denying the idea. The journal entry does at least indicate that he was beginning to explore new territories and even entertain ideas that were then scientifically heretical.

A second interesting point about this statement is the echo of the comment about the giant tortoises made by Nicholas Lawson, the Galápagos Islands' governor. In time the importance of this single throwaway remark would grow in stature until it dominated Darwin's life – but more of that later.

Once back at home in England, Darwin was able to reflect on the many things he had learnt from his time on the *Beagle*. He looked back on the giant armadillo and sloth fossils from the South American pampas and marvelled at how these extinct giants had been replaced by their smaller, modern equivalents. Surely, he thought, was this not evidence of a law of succession in the fossil record?

Then there were the Galápagos mockingbirds with their separate 'varieties', each occurring on a different island. If these proved to be species and not varieties, could the birds have once belonged to the same species whose form had been changed by the isolation of island life?

As he rested at his father's sumptuous country home in Shropshire, turning these thoughts over in his mind, the young naturalist knew that his most important task was not to pursue the theory of evolution but to make his name known within the closed circle of Victorian science. The best way to do that was to get his findings from his time on the *Beagle* into print – and quickly. FitzRoy had already approached Darwin with the idea of producing a joint narrative of the voyage in which the captain would write about the seafaring and technical details while the naturalist would handle the animals, plants and geology.

To achieve this Darwin would have to disseminate his collection of fossils, plants and animals to various experts around the country. Apart from the fact that he had neither the time nor the expertise to study himself all the things that he had collected, professional opinions would give much-needed weight to any published work. Barely more than a month after returning from his travels Darwin found himself touring the country, in search of the requisite experts.

Given the extraordinary nature of his voyage, he found no shortage of willing takers for his precious specimens. The list of experts, which included names such as Richard Owen, William Buckland and George Waterhouse, numbered many who were at the head of their field and whom Darwin revered. Now he was mixing with them.

Richard Owen was so impressed with the South American fossils that he asked for further specimens to be sent to him post-haste. Even Darwin's hero Charles Lyell, the author of his much-prized *Principles of Geology*, took an interest in his observations on volcanoes and coral islands. It was a giddy time and for a while the young man leant towards becoming a geologist rather than a zoologist. Then, around Easter 1837, he came down to earth with a bump.

During Darwin's first round of specimen touting, Thomas Bell, a recently appointed professor of zoology at King's College, London, had agreed to look at the reptile specimens and was now ready to give his opinion on them.

Amid the qualified observations on the various lizards and snakes that Darwin had picked up from various countries, Bell also gave his opinion on the Galápagos tortoises. These, he pronounced, were almost certainly native to the Galápagos Islands themselves and had not, as most people believed, been taken there from the Indian Ocean by buccaneers, sailors and seafarers. Prior to Bell's pronouncement, Darwin had no cause to think that the tortoises of the Galápagos were anything other than immigrants from the Mascarenes. In fairness, neither did anyone else.

Even before his visit to the islands, Darwin knew that all the giant tortoise species in the world (that is, those from the Indian Ocean and the Galápagos) were classified under Schweigger's name of *Testudo gigantea*. While at the Galápagos, FitzRoy had also told Darwin that these creatures were likely to have been imported.

'There is no other animal in the whole of creation,' wrote FitzRoy later, 'so easily caught, so portable, requiring so little food for a long period, and at the same time so likely to have been carried, for food, by the aborigines who probably visited the Galápagos Islands on their balsas [rafts], or in large double canoes, long before Columbus saw that twinkling light, which, to his mind, was as the keystone to an arch.'

Like his comrades, Darwin subscribed to this view, and as he travelled from one Galápagos island to another he assumed that all the tortoises he encountered were from the same species, *Testudo gigantea*. He therefore did not pay as much attention to them as perhaps he could have done.

On the assumption that before him was the already copiously described Mauritius tortoise, Darwin had focused on the animals' behaviour and not their appearance. He had made plenty of notes concerning their feeding and breeding habits and, of course, what they tasted like, but very few on their physical characteristics. He also did not attempt to collect them although, in fairness, an adult tortoise's shell (even when empty) would have been prohibitively large and heavy to be stored long-term on a small ship such as the *Beagle*.

Thomas Bell's belief that the giant tortoises were native to the Galápagos meant that they were probably a different species from the *Testudo gigantea* of the Indian Ocean. The Galápagos, it would appear, had its own as yet unnamed species of giant tortoise but did Bell's pronouncement have further implications?

As Darwin considered this, the words of Governor Lawson came floating back to him. 'Show me a tortoise,' Lawson had said, 'and I'll tell you which island it came from.' This implied that, like his mockingbirds, there could be several tortoise species in the Galápagos, each belonging to a different island.

If each Galápagos tortoise species were restricted to a specific island, surely the only reasonable explanation was that their isolation from one another had somehow caused them to evolve into new species. There was no doubt at all that the tortoises on the Galápagos were related to one another and that they shared a common ancestor – Darwin had seen that for himself – but if what Bell and Governor Lawson were suggesting was true, it appeared that descendants of that common tortoise ancestor had become stranded on discrete islands and that this very isolation had led to their evolving into different species.

The stark implication of this was that if members of the same species became isolated from one another, each population would start to evolve in different ways until a completely new species of animal resulted. The Galápagos Islands appeared to be offering Darwin proof that the controversial notion of evolution was a reality. At last Darwin was to have his eureka moment. Unfortunately, it had come more than eighteen months too late.

Reflecting on the words of Governor Lawson, Darwin knew that the only way to test the idea that different tortoise species were indeed restricted to different islands was to examine several specimens whose original island localities were known, and to compare them side by side. Only then could he know for sure whether or not any differences were restricted to a particular island.

While it was an excellent idea in principle, Darwin looked in dismay at his lone pet tortoise, collected from San Salvador Island. Despite his having visited four of the Galápagos Islands, this one juvenile specimen was all that he had to show for his trouble.

Worse still, at one point the *Beagle* had had over thirty adult tortoises on board, collected by FitzRoy as a source of food for the Pacific crossing. Had Darwin realised earlier the significance of Lawson's words, he could have saved the remains of the slaughtered tortoises. Instead he had watched silently as each tortoise was carefully selected, butchered and cooked during the Pacific voyage. After having eaten the tortoise, Darwin and the rest of the crew had stood by as the shell, bones and other inedible fragments were brought up from the galley and unceremoniously tipped over the side into the deep waters below. Ironically he had managed to eat his way through the most important specimens on board the *Beagle*. This was to be the first in a series of disheartening jolts for Darwin.

There was one last hope. In addition to his own pet tortoise, Darwin also knew that Captain FitzRoy had taken two from Española Island, while Syms Covington had another from San Salvador Island. Four tortoises from three different islands: all were still available to him and, by good fortune, were located in London. It was a last-ditch attempt but worth a try.

Thus it was that a few weeks later Darwin stood inside the hallowed halls of the British Museum with John Gray, the resident reptile expert, and four small Galápagos tortoises only a few centimetres in length. The young naturalist looked on expectantly as his learned peer picked up each animal and slowly turned it over in his hands, searching for the tell-tale characteristics that would allow him to subdivide the animals into two or more different species. At last Gray was ready to make his pronouncement.

First of all he corrected Darwin on his use of scientific names for the giant tortoises. After Darwin had left England in 1831, John Gray had superimposed his own name for the world's giant tortoises on that of Schweigger, coined in 1812. They were no longer *Testudo gigantea* as Schweigger had decreed but *Testudo indica* as laid down by Gray in 1831 in his all-encompassing *Synopsis Reptilium*. The renaming was simply a piece of scientific one-upmanship. Gray still believed that *Testudo indica* was a single species that covered the tortoises of both the Indian Ocean and the Galápagos Islands. Perhaps, Darwin hoped, that was about to change. What was Gray's view on his new specimens? Was there more than one species of tortoise living in the Galápagos?

Gray delivered his opinion but the news was not favourable. Shaking his head, he revealed to Darwin that the tortoises were too immature to be able to see any differences in their shell shape. At such a young age, Gray explained, all tortoises look alike. Wait another two or three decades, he advised, and then it might be possible to see the differences in shell shape, colouration, etc. To the impatient Darwin this was no good at all – it would be quicker to return to the Galápagos and collect more tortoises. Even the museum's own collections were of no help. The few Galápagos tortoise specimens in their possession had been donated by ships and their island of origin was uncertain.

Doubtless the thought of all those empty tortoise shells lying around Santa María Island, being used as plant pots or just left to rot, haunted Darwin. Why had he not collected at least one of them? Worse still was the image of the remains of all those cooked tortoises from the *Beagle*, now lying under several thousand metres of water at the bottom of the Pacific Ocean. Any ideas he had of testing Governor Lawson's assertion about the Galápagos tortoises now appeared to be entirely lost. All Darwin had as proof that evolutionary processes were operating in the Galápagos were his mockingbird specimens, which were so few in number that the discovery that there were different varieties living on different islands could be merely a coincidence or the result of insufficient specimen material.

All thoughts of the Galápagos tortoises were temporarily put to one side as Darwin continued to draw together the information that would eventually form his zoological volume on the *Beagle's* voyage. The issue of evolution was not to desert him for long. At last, luck was on his side.

～

On 4 January 1837, Darwin visited the headquarters of the Zoological Society of London, bringing with him the several hundred mammal and bird specimens that had been collected during his overseas adventure. As the Society had scheduled a council meeting for that day, a large number of eminent zoologists were wandering about the hallowed corridors of the Regent's Park building. One by one Darwin managed to get them interested in his specimens, many of which were obviously completely new to science. Soon offers of help were coming from all directions.

The zoologists James Reid and William Martin split up the eighty-odd mammal specimens between them, with Martin taking the cats and Reid the marsupials and large quadrupeds.

The bird specimens were more difficult. As there were over 450 of these, sorting through them would be an onerous task and there were few people with enough time on their hands or experience of foreign birds to undertake it.

In fact, by far the best candidate was John Gould, the Zoological Society's own resident ornithologist. Gould was only a year older than Darwin but he was already an authority on birds of all kinds, especially those from exotic lands. He was also an acknowledged genius when it came to identifying anything with a feather attached to it and a talented artist to boot.

In the absence of any other offers, Darwin handed over his entire bird collection to John Gould in the hope that he could try to make sense of them. This turned out to be a stroke of luck that would more than make up for the lost opportunities of the eaten and discarded tortoises.

Gould, who was still fired with youthful energy, instantly gave his undivided attention to Darwin's new and tenderly preserved specimens. This was perhaps as well because although the young Darwin's general knowledge of the natural world was impressive, his ability to separate his animal specimens into their various species was not so acute. During his time on the *Beagle* he had been particularly inept at identifying the hundreds of bird specimens he collected, especially when confronted with new and unfamiliar species. Faced with over 400 specimens and Darwin's many identification errors, Gould certainly had his work cut out.

He moved fast, relishing the challenge of handling birds from parts of the world that were new to him. After a cursory glance (and possibly after some prompting by Darwin), Gould recognised that the Galápagos birds were by far the most intriguing and interesting in Darwin's vast collection. He went straight to work on them.

Within only a few days he had focused on the Galápagos finches. Darwin had had a great deal of trouble identifying these birds but because of his expertise, Gould could immediately see where the naturalist had gone wrong. Less than six days after receiving Darwin's birds, he was able to stand up at a Zoological Society meeting and announce that he had studied a 'series of ground finches, so peculiar in form that I am induced to regard them as constituting an entirely new group containing fourteen species, and this appears to be strictly confined to the Galápagos islands'.

On the evening of Gould's announcement Darwin was in Cambridge but was delighted to see both his name and his bird specimens featured in the London *Morning Herald* and several other newspapers, all of which acknowledged the scientific value of his collection. The five years of discomfort, homesickness and seasickness endured on the *Beagle* were suddenly beginning to bring rewards.

In the coming weeks Darwin's specimens were discussed again and again at the Zoological Society's meetings but the collector's absence created difficulties. Gould, who kept uncovering new and interesting bird species, badgered the continually missing Darwin for further information about the specimens, but only so much could be conveyed by letter. 'I cannot enter into any further details respecting the Galápagos finches,' announced Gould at one meeting, 'until Mr. Darwin has furnished me with some information relating to their habits and manners.'

It became clear to Darwin that if he was to make headway in his desire to be a naturalist, he would have to move from Cambridge to

London as that was where the hub of England's naturalists was located. Darwin had no love of the city, finding it claustrophobic, smelly and uncultured, but, in early March 1837, he arrived at his brother's house in central London with the intention of staying there permanently. However unpleasant the city was, it had its uses, and only a couple of days after his arrival Darwin engineered a meeting with John Gould, anxious to hear his news about the bird specimens. It was better than he could have dared hope.

Gould and Darwin sat facing each other across a table in the young ornithologist's Zoological Society laboratory. Around them were Darwin's beautiful but lifeless bird specimens, neatly divided into separate piles.

Gould's first pronouncement was his most dramatic, but he almost seemed to apologise for it. He confessed to Darwin that almost all of his and FitzRoy's original identifications had been discarded and new names given in their place. Nowhere was this more true than with the Galápagos specimens. The nature of these birds, Gould admitted, had taken him quite by surprise.

Of the twenty-six species of land bird that Darwin had collected from the Galápagos, said Gould, twenty-five were completely new to science and could be found nowhere else in the world. (The odd one out was the American bobolink, a very wide-ranging bird.) Even the seabirds, many of which could fly long distances, were also unique to the Galápagos. The archipelago had proved to be something of an ornithologist's paradise.

As Gould spoke, Darwin scribbled away, frantically trying to keep up as Latin names and specimen numbers were flung across the table at him in quick succession. Not realising the importance of his specimens, Darwin had only brought one sheet of paper with him and as Gould continued to throw scientific gems at him, so Darwin's spidery writing became tighter and more frantic. The last notes had to be squeezed in sideways at the bottom of the page.

When Gould had finished, the excited Darwin asked, not for the first time, whether he was sure about his identifications. Could not the mockingbirds, for example, just be varieties and not distinctive species? Gould looked back at his colleague. 'If they are varieties,' he replied, 'then the experience of all the best ornithologists must be given up.' This was good enough for Darwin who remained silent, the excitement growing within him. Although Gould did not realise it, every piece of information that he was handing over was contributing to Darwin's increasingly positive view of the 'instability of species'.

The news that all of Darwin's Galápagos finches were entirely new to science was one thing, but it was Gould's confirmation of Darwin's suspicions about the mockingbirds that was the real revelation.

The previous summer, Darwin had noted that the various mockingbird species seemed to occur on different islands. Ever unsure about his own ability to identify species, Darwin had suspected that the mockingbirds were simply 'varieties' and not separate species. Were they proved to be species, Darwin had written, then 'such facts would undermine the stability of species'.

Now, several months later, Gould sat before him, announcing that three of the four 'varieties' of mockingbird that Darwin had identified were indeed separate species, each being found only on its own discrete island. The words echoed through Darwin's head, triggering memories of Bell's findings on the tortoises and Governor Lawson's boast, but the icing on the cake was yet to come.

In his overall summary of the Galápagos, Gould commented that most of the birds there appeared to be related to similar species found on the South American continent. This included the finches and the mockingbirds, both with distinct, but clearly related, species on the mainland.

To Gould this was just a point of interest. Like Charles Lyell and others, he saw nothing subversive in the strange genealogical links between separate animal populations. To Darwin, however, this news was dynamite, for the Galápagos were not the only remote island chain he had visited. Early in the *Beagle*'s voyage he had also called at the Cape Verde Islands, lying just off the West African coast. Like the Galápagos, they are volcanic, sparsely populated and located in a cold-water ocean current. In terms of climate, situation, geology and topography, the Cape Verde Islands are very similar to the Galápagos; they too had their own range of endemic birds, reptiles and mammals, which also seemed to be related to species on the nearby African mainland.

Endemic Galápagos bird species of South American origin and endemic Cape Verde bird species of African origin: it looked very much as if, when these sterile volcanic islands emerged from the sea, they were invaded by animals from the adjacent mainland. Having become isolated on these islands, the descendants of these mainland species had transformed themselves into new species. Even an expanse of water stretching a few tens of kilometres between individual islands seemed to be enough to produce different species on different islands.

The meeting with Gould ended and as Darwin headed back into central London his head was swimming with all that he had heard. There could now be no doubt about the importance of Governor Lawson's comment that each Galápagos island had its own variety of tortoise.

Lawson's remark, as we have seen, was made to Darwin during a long day when he was party to many other conversations. Why this passing boast from a colonial official should have stuck in his mind when many other snippets of information must have escaped his memory we shall never know. I suspect that even Darwin himself would have been hard pushed to explain why this observation echoed in his brain but in the days after his meeting with Gould he would cogitate over it again and again.

Darwin had mentioned Lawson's words in his zoological notebook of September 1835, written while he was in the Galápagos. Here, in among a long list of scribbled notes on the tortoises, Darwin simply says: 'Mr Lawson states he can, on seeing a tortoise, pronounce with certainty from which island it has been brought.'

Later, in the second (1845) edition of his book about his time on the *Beagle*, Darwin would write: 'I did not for some time pay sufficient attention to this statement, and I had already partially mingled together the collections from two of the islands.'

Thanks to John Gould's extraordinary abilities, in less than a day Darwin had gone from being a sceptical Creationist to a reluctant evolutionist. His as yet ill-formed ideas seemed to have been given substance and yet there was still a long way to go before his suspicions could be proved beyond doubt. To do this was going to require cunning and skill.

As things stood, Darwin had only the mockingbirds with which to make his case but, to ensure that he was not mistaken, he would need to find other Galápagos species that were unquestionably restricted to individual islands.

The ideal candidates, the tortoises, had all been eaten so he turned to his other specimens. The mammals of the Galápagos were of no use: not only were they few in number but, as far as Darwin could see, they had almost all been imported into the islands by humans. The reptiles were also of no use. This left only the tortoises and the iguanas, and Darwin had not collected enough of either to be certain. In short, he had only his birds, plants, fish and insects to choose from.

The insects had been given to George Waterhouse at the British Museum but they too were few in number and, like the finches, mostly without locality information. Waterhouse was a busy man and it would be some months before he would report back to Darwin. Gould was already dealing with the birds, which only left the plants and the fish. If his theory was to be proved correct, he would need to get somebody to study these specimens, and quickly.

Darwin dashed off a letter to his Cambridge mentor John Henslow, asking him to look at the Galápagos plant specimens that he had handed over a few months before. At the same time he asked Henslow to get somebody to look at his Galápagos fish specimens. All of this, said Darwin, was needed with some urgency so that he could include the studies in his published résumé of the *Beagle* voyage.

The letter was carefully worded so as not to rouse suspicion. In an age where evolutionary scientists were held in much the same esteem as prostitutes and lepers, Darwin was careful to keep hidden his real interest in the Galápagos plants and animals.

While he waited for others to report back to him, he would have to do some homework himself and where better to start than with the birds?

Although the mockingbirds had provided the inspiration for his theories, the most numerous endemic bird species in Darwin's collection were those of the finches. Of the thirteen species that Gould had identified, all were closely related to one another and also, distantly, to the South American finches. More importantly, none of the Galápagos species were found anywhere else in the world. If Darwin could prove that all or some of these species were restricted to individual islands, the case in favour of evolution would be proved. So it was that Darwin returned to the Zoological Society with the intention of studying his finch specimens.

However, he was in for a big disappointment for not a single one was labelled with its island of origin. How on earth was he supposed to tell if an individual species was restricted to one island if he could not tell where *any* of the specimens had come from? Even looking at his notebook did not help. He had started labelling his Galápagos specimens weeks after the *Beagle* had left the archipelago, by which time the exact locality for each one had been forgotten. The finch specimens were simply labelled as having come from the Galápagos.

If proof were ever needed that thoughts of evolution were far from Darwin's mind whilst he was resident in the islands, this was it. Even after Governor Lawson's boast Darwin had still not seen the implication of different animal species on different islands. It was only by good fortune that his servant had noted the mockingbirds' islands of origin. If he had not, Darwin might never have seen the connection between endemism and evolution but, even so, without further physical evidence that connection would be difficult to prove.

Darwin racked his brains for ways in which the original localities of the finch specimens could be reconstructed. He knew that he had collected finches from only three of the four islands he had visited. The omission of the fourth island, Isabela, was not from lack of opportunity but laziness.

As he checked through his notes Darwin found that while on Isabela he had stood by a small puddle of water and watched as 'doves and finches swarmed around its margin'. Dozens of potential specimens had been within the young naturalist's grasp and yet he had not lifted a finger to capture them. Like the tortoises, this was another mistake that was now costing Darwin dear. Even if he could find a way of identifying which island each specimen had come from, would birds from only three islands be enough to make his case?

Fortunately there was a possible solution to the problem. Although he was the *Beagle*'s official naturalist, while on board Darwin had actually been surrounded by several amateur naturalists, some of whom were also assembling their own personal collections of plants and animals. If some of these collections contained Galápagos finches and if they were labelled with their island of origin, Darwin might be able to work out which islands his species had come from. As it happened, he was in luck.

The week before Darwin had moved to London, Captain FitzRoy, perhaps spurred by the publicity being given to his young naturalist's specimens, had himself visited the British Museum and deposited there a portion of his own animal and plant collection. This included thirteen finch specimens from the Galápagos, all of which were fully labelled with their island of origin.

To Darwin this was a double stroke of luck, as it meant that not only were there further specimens to examine but also that he would not have to pay a visit to the increasingly fundamentalist FitzRoy, who was beginning to see the work of the Devil in every corner of society. A brief letter to FitzRoy asking for any further information about the finches was all that was needed. It was not even answered by FitzRoy himself but by one of his secretaries.

As well as FitzRoy's birds, Darwin knew of two other collections. One was held by his servant on the *Beagle*, Syms Covington. Darwin had trained Covington in how to collect and preserve specimens and the young man had become so enthusiastic that he began collecting for himself. Fortunately Darwin had retained the services of Covington after the *Beagle*'s return and he was pleased to find that among his servant's birds were four Galápagos finches from two of the islands.

The last collection was that of Harry Fuller, the *Beagle*'s steward, who for a week had helped Darwin shoot and preserve birds on San Salvador Island. He had eight finch specimens, labelled as coming from San Cristóbal and San Salvador islands.

Darwin hoped that these might offer him the proof he sought. There was also the chance that, by comparing them to his own collection, he might just be able to reconstruct from which island each of his own birds came.

Of his thirty-one finch specimens, Darwin's notes allowed him to identify only three (all of the same species) with an individual island.

The rest would have to be worked out by comparison to his ship-mates' collections or by educated guesswork. He began this task as soon as time permitted but the results produced a confused picture. Although it superficially appeared that the finch species were indeed endemic to particular islands, there was so much uncertainty over the locality data that it would be easy for others to discredit his evidence. The finches provided enough corroboration for Darwin to be able to convince himself, but nowhere near enough to convince his scientific colleagues.

By the summer of 1837 he was under pressure to finish the Galápagos chapter for his volume on the zoology of the *Beagle* voyage. Although the locality information on his finches was far from satisfactory, it would have to do.

At the end of all this Darwin was still left with a problem. An insufficiency of specimens and poor labelling had left him no nearer to proving absolutely whether or not evolutionary processes were occurring on the Galápagos. Nonetheless, the evidence from the tortoises, finches and mockingbirds all appeared to lend weight to the idea that animal and plant species were capable of evolving into entirely new species.

In July 1837, after months of serious thought, Darwin finally took the plunge and cut the link to his Creationist past. In that month he began a new notebook in which to record his scientific ideas. It was entitled 'Transmutation of Species'. In his private diary Darwin recorded the event: 'Opened the first notebook on "Transmutation of Species" – Had been greatly struck from about March on character of South American fossils and species on Galápagos Archipelago. These facts [are the] origin (especially latter) of all my views.'

Darwin's conversion to the evolutionary cause was now complete but he was far from finished with the Galápagos Islands and especially their tortoises. The idea that there were different species of tortoise restricted to particular islands obsessed him. The evidence from the birds had proved to be useful but was open to argument. The words of Governor Lawson made it sound as though the case for the role of evolution in the development of the Galápagos tortoises was much more definite. If an amateur like Lawson could tell the island tortoise races apart, surely this was a clear-cut case of evolution at work.

The problem continued to haunt Darwin but, without more specimens, any further progress on the matter looked unlikely. The time had come for him to write up the section on tortoises for his publishers and yet the hoped-for evidence to back up Lawson's assertion was still not there. When Darwin came to transcribe behavioural observations, he added only general comments about their speciation:

> It was confidently asserted, that the tortoises coming from different islands in the archipelago were slightly different in form; and that in certain islands they attained a larger average size than in others. Mr Lawson maintained that he could at once tell from which island any one was brought. Unfortunately, the specimens which came home in the *Beagle* were too small to institute any certain comparison. This tortoise, which goes by the name of *Testudo indicus*, is at present found in many parts of the world. It is the opinion of Mr. Bell, and some others who have studied reptiles, that it is not improbable that they all originally came from this archipelago. When it is known how long these islands have been frequented by the buccaneers, and that they constantly took away numbers of these animals alive, it seems very probable that they should have distributed them in different parts of the world. If this tortoise does not originally come from these islands, it is a remarkable anomaly; inasmuch as nearly all the other land inhabitants seem to have had their birthplace here.

To the casual reader Darwin was merely relating interesting titbits about the distribution of the Galápagos tortoise. In reality, he was secretly conveying his new evolutionary beliefs to an unwitting audience. Different tortoises on separate islands and yet they were all indigenous. It was as if Darwin was writing in code, waiting for others to make the right connections so that they might crack it for themselves. It would be decades before others would understand exactly what he was driving at.

In his original description Darwin was careful not to mention whether the differences between the Galápagos tortoises were due to there being a great variety in one species or because there were several distinct species. This was because Darwin himself did not know. For him it was the question of the moment and yet, short of returning to the Galápagos, there would be no way of telling. Or so he thought.

By February 1838 the proofs for his *Beagle* journal were complete and his tortoise notes set in stone. Evolutionary concepts were still at the front of his mind. His first transmutation notebook had already been filled with scribbled ideas and jottings, and a second was already well under way.

It was at the end of February that Darwin made his usual bi-weekly trip to the Zoological Society to listen to that evening's speakers. On the bill that night was the young French reptile expert, Gabriel Bibron, who, together with Marie Constance Duméril, had recently finished compiling a wide-ranging natural history of the world's reptiles. His report that evening was equally wide-ranging and included interesting snippets of information that their research had turned up.

As Bibron's talk moved on to the subject of the land tortoises, Darwin was all attention. Bibron began by describing his recent studies on material from the 'Isle de France' (Mauritius) where he had looked at fossilised material from both the giant tortoise and the dodo. After due consideration, Bibron had concluded that the Indian Ocean contained at least three species of tortoise: *Testudo elephantina* from Aldabra, *Testudo daudinii* from the Seychelles and *Testudo peltastes* from the Mascarene Islands. That caused Darwin to sit up and take notice but better news was to come!

With complete confidence Bibron announced that he was certain that the Galápagos tortoise was of a completely different species to that of the Indian Ocean. As such he and his colleague had now classified the Galápagos tortoises under the name *Testudo nigrita*. Instead of one just giant tortoise species, the world now had four: three in the Indian Ocean, one in the Galápagos. (It should be noted that, true to form, Duméril and Bibron ignored Schweigger's *Testudo gigantea* and Gray's *Testudo indica*, overwriting them with their own classification, *Testudo elephantina*. It would be up to future researchers to sort out this taxonomic mess but in his published works Darwin continued to use Gray's *T. indica*.)

Not only did this validate Thomas Bell's assertion that the tortoises Darwin had seen were native to the Galápagos but it also meant that the idea of Mauritius being populated by Galápagos tortoises that had been taken there by ship was false. This was to add even more weight to Darwin's growing conjectures on evolution. According to Bibron, the four known species of giant tortoise were all to be found on remote islands, reinforcing Darwin's observed link between isolation and evolution.

After the meeting an excited Darwin approached Bibron, anxious to find out more. In turn, Bibron was just as keen to hear about Darwin's voyage and all the strange reptiles that he had encountered on the way. The two men talked, swapping information, discussing their various scientific discoveries, but it was to be Darwin who would get the most out of the discussion.

Although Bibron had not been to the Galápagos, he had a particular interest in the islands' tortoises. When he made his initial study of the giant tortoises, the Paris Natural History Museum had only one Galápagos juvenile tortoise specimen but it had recently obtained more.

Darwin took a risk with Bibron and voiced his thought that there could be great variation among the Galápagos tortoises on different islands. The governor had even said as much but Darwin confessed that he had not had the opportunity to check this for himself. Could Bibron shed any light on the matter?

Darwin held his breath as Bibron, who was his equal in age and status, absorbed the question, formulated a reply and mentally translated it into English. 'Yes,' said Bibron, 'I believe that there are at least two distinct species of tortoise from the Galápagos Islands. Whether they are restricted to separate islands, I cannot say.'

Darwin was overjoyed. Even though Bibron could not say for certain that there were different tortoise species on particular islands, his identification of two species seemed to back up Governor Lawson's idea that each island had distinct tortoise species. It seemed to Darwin that what had once been a vague notion based on three mockingbird species was gaining serious credence. He now had firm evidence from the mockingbirds, and circumstantial evidence from the finches and tortoises.

Darwin was now more convinced than ever of the reality of evolution: that one species could transmute itself into another by natural means alone. Journeying home from the Zoological Society, Darwin was elated. Later he wrote excitedly in his notebook: 'The

French Bibron says that two species of tortoise come from the Galápagos!!!'

Bibron's information and his multitude of new species names had come too late for Darwin to include in the main text of his journal, which was already in production, but he could at least include the material in the footnotes. The first seeds of what would eventually be used to bolster his ideas about evolution were dashed off and sent to the publishers. Only Darwin knew their real significance.

> There is every reason for believing that several of the islands possess their own peculiar varieties or species of tortoise, but that my specimens were too small to decide this question. M. Bibron now informs me, that he has seen full-grown animals, brought from this Archipelago, which he considers undoubtedly to be distinct species . . . Doubtless the several islands have their own representatives of the [marine iguana], like they have of some of the birds, and of the tortoises.

By the spring of 1838, Darwin's volume for the *Journal of Researches* was safely with the publishers. Given that it was barely eighteen months since his return, he had done a remarkable job. Not only had he written an entire volume of several hundred pages, but he had also cajoled some of the most eminent scientists in the land into looking at his animal, plant and fossil specimens. He had even managed to partially snatch victory from the jaws of defeat, finding inventive and ingenious ways around the poor labelling of his Galápagos specimens.

In truth, however, Darwin was no nearer to understanding the significance of his evolutionary discovery. He was not the first person to conclude that species must somehow spontaneously transmute; the idea had been around for decades. The problem was that nobody could yet satisfactorily explain how and why such changes happened. Nevertheless, there were more difficulties with the concept of evolution than just its inexplicability, and Darwin was about to witness this at first hand.

In the autumn of 1838, Robert Grant, one of Darwin's old tutors from Edinburgh who was now working for University College,

London (the so-called 'Godless college'), dared to criticise the estab-lished view that the earliest fossil mammals were to be found in the ancient Jurassic rocks of Oxfordshire. Grant was a fanatical believer in evolution and saw life on earth as having progressed from lower forms, such as fish, through intermediate forms, such as reptiles, to higher forms, such as mammals. He could not therefore entertain the idea that the advanced mammals could have existed as far back as the Jurassic period, which lay in the 'age of reptiles'.

Unfortunately, Grant chose to make a stand over a small fossilised jawbone from the Jurassic, which he insisted was from a reptile but which almost everybody else, including Darwin, believed to be from a primitive mammal. Grant's vocal opposition was just what his ene-mies had been waiting for. At a meeting of the Geological Society on 19 December 1838, Darwin watched as his old mentor was pulled to pieces in a pre-organised and multi-modal attack by such scien-tific heavyweights as Richard Owen and William Buckland. By the end of it Grant looked like a thoroughly amateur scientist who had let his belief in evolution bias his opinion on the ancient mammal fossil.

It was a swift lesson to Darwin that any notions concerning evo-lution were not at all welcome in higher circles of the scientific estab-lishment. There, the idea of an evolutionary progression in the fossil record not only went against the Bible but also denied man his exalted place above the animals. The ritual humiliation of Grant told Darwin that if he were to establish his reputation as a naturalist, he would have to keep his evolutionary ideas to himself.

This was unfortunate for, around the same time that Grant was being attacked, Darwin was actually beginning to determine which processes were responsible for evolution.

Some weeks earlier, he had chanced upon a copy of the recently deceased Reverend Thomas Malthus' *Essay on the Principle of Population*, a work of economics, which dealt with society's eternal problem of the struggle for resources. He was particularly obsessed by the human population's natural ability to grow faster than its abil-ity to feed itself. Malthus saw society as being in a continual war for limited resources. Those who could get enough would survive; those who could not would fall by the wayside.

Darwin was no fan of Malthus, but as he digested his ideas, parallels began to emerge between the Malthusian view of society and the way in which the natural world operated. Did not animal and plant species also compete for limited resources? Would not individuals that could not obtain enough resources die?

The seeds of what would eventually become Darwin's theory of natural selection had been sown. Like Malthus' view of society, Darwin envisaged a natural world where all organisms competed for a limited number of resources and where even the slightest advantage might be the difference between survival or death. Those animals that did survive would be more likely to breed, passing on their physical characteristics to their offspring. The variation that Darwin had seen within a single species made him think that if an individual animal was born with a physical characteristic (for example, bigger teeth) that gave it an advantage, however small, over its rivals, this animal would be more likely to survive and breed, thus passing on this characteristic to the next generation. Over thousands of generations the accumulation of many such small physical characteristics would be enough to create an entirely new species. If the same species became separated into several distinctive populations (for example, with habitats on different islands), with time the physical changes would be slightly different in each case, producing a new and distinctive species for each population.

The discovery of an explanation as to how a species might evolve over time instantly put Darwin in a class of his own. The stumbling block for people such as Robert Grant had always been their inability to explain their beliefs. Darwin had overcome that hurdle but he knew that he would have to remain silent about his findings as he still did not have enough evidence to justify his belief in evolution. If a respected zoologist like Grant could be destroyed in public, the establishment would not think twice about demolishing a young whippersnapper like Darwin.

Darwin silently accepted his vow of silence but at the same time determined to gain the evidence he needed and so, as his volume on his *Beagle* findings was published to great acclaim, he began to look for the voluminous amount of evidence necessary to prove his theory valid.

In the short term this search would move away from the *Beagle* specimens into other fields, such as domestic dog and pigeon breeding, but the Galápagos would not be forgotten.

~

The passage of several years saw a Darwin who was very different from the young man who stepped off the *Beagle* in 1835. In 1843 he was married, had children and was living on a generous allowance from his father in a large house in Kent that was rapidly becoming a laboratory devoted to the cause of evolution.

His book, *Journal of Researches*, published in 1839, had been a great success, far more so than FitzRoy's own volume, and the publishers now wanted a revised edition. This meant tying up a few loose ends and at the top of the list were the Galápagos plants and tortoises.

After much coaxing, his old Cambridge professor had still not managed to get very far with the Galápagos plants, although he did pronounce that 'there are several instances of distinct species of the same genus, sent from one island only: that is, whilst the genus is common to two or three islands, the species are often different in the different islands. In some cases the species seem to run very close to each other, but are, I believe, distinct.'

The plants were one of the few groups of specimens for which Darwin had good location information. This is ironic as, of all the wonders in the natural world, he was least interested in the plants and was particularly maladept at telling one species from another. Nonetheless, he had diligently collected a good many specimens from the Galápagos and, because they needed immediate pressing, had labelled them correctly.

They were now Darwin's best hope but finding somebody to give an expert opinion on them had proved difficult. Fortune dictated that he came into contact with Joseph Hooker, a young botanist who had just returned from an Antarctic voyage. Although the two men were not well acquainted, in 1843 Darwin wrote to Hooker, wondering if he would be interested in examining his plant specimens from Tierra del Fuego, which might correspond to his own southern hemisphere collection. As an afterthought Darwin suggested that Hooker might look at his Galápagos plants, adding that the 'flora of this archipelago would, I suspect, offer a nearly parallel case to that of St. Helena, which has so long excited interest'.

Hooker took the bait and agreed to inspect the plants. It was to be the beginning of a lifelong friendship, the opening months of which consisted of a continuous stream of letters from Darwin to Hooker, asking questions about the Galápagos plants. Were the species separate to each island? Were they related to any species on the mainland?

Finally, in 1844, Darwin was to hear the result of Hooker's work. Yes, he could confirm that there were different species to be found on separate islands, to an extreme degree.

Of the thirty-eight species on San Salvador Island, thirty were found nowhere else. On Isabela it was twenty-two out of twenty-six; on San Cristóbal it was twelve out of sixteen; and on Santa María Island twenty-one out of twenty-nine.

This time there could be no doubt: each Galápagos island did have its own unique group of plant species, fitting into Darwin's scheme perfectly. Darwin greeted the news with joy and relief, writing back to Hooker: 'I cannot tell you how delighted and astonished I am at the results of your examination; how wonderfully they support my assertion on the differences in the animals of the different islands, about which I have always been fearful.'

Not long after Hooker's plant conclusions came George Waterhouse's insect results. The insects had been low in number, only a few having information on location, but nonetheless Waterhouse noted that of the species with a given locality, no two were the same.

It was nearly a decade since Darwin had visited the desolate and craggy Galápagos Islands and only now was he reaping the rewards. He at last had the satisfaction of knowing that he was right, even if he could not yet announce his findings to the world.

All of this good news arrived just in time for Darwin to include it in the revised edition of his *Journal of Researches*. The slim footnote of his 1839 edition was now expanded to over 2,000 words. The tentative language of the first edition was gone; in its place were firm assertions about the unique nature of the islands' wildlife.

I have not as yet noted by far the most remarkable feature in the natural history of this archipelago; it is, that the different islands to a considerable extent are inhabited by a different set of beings . . . I never dreamed that islands, about 50 or 60 miles apart, and most of them in sight of each other, formed of precisely the same rocks, placed under a quite similar climate, rising to a nearly equal height, would have been differently tenanted; but we shall soon see that this is the case. It is the fate of most voyagers, no sooner to discover what is most interesting in any locality, than they are hurried from it; but I ought, perhaps, to be thankful that I obtained sufficient materials to establish this most remarkable fact in the distribution of organic beings.

Darwin listed the results of a decade's work with ease. The reader could scarcely guess the secret that he was holding nor the anguish that gathering the Galápagos data had caused him. Not once did he suggest why the different islands should have different species, let alone dare to stray into language that might be considered pro-evolution in nature. Indeed, when summing up Darwin even disguises his true beliefs by using Creationist language:

Reviewing the facts here given, one is astonished at the amount of *creative force*, if such an expression may be used, displayed on these small, barren, and rocky islands; and still more so, at its diverse yet analogous action on points so near each other. I have said that the Galápagos Archipelago might be called a satellite attached to America, but it should rather be called a group of satellites, physically similar, organically distinct, yet intimately related to each other, and all related in a marked, though much lesser degree, to the great American continent.

The loose ends tied up, Darwin could now leave behind the issue of the Galápagos Islands and concentrate on gathering more concrete evidence in favour of evolution through natural selection. Having dominated his thoughts for so long, the significance of the islands and their wildlife would diminish as better and more conclusive data emerged in favour of natural selection. When, in 1859, Darwin published his theory in the now legendary *Origin of Species*, the Galápagos Islands warranted only two mentions and then only Hooker's plants get any real coverage. Darwin's collecting methods meant that the data from the tortoises, finches, insects and other species were too suspect to be trusted and, in the glare of publicity that *Origin* would receive, too dangerous to risk using. Nonetheless the tortoises of the Galápagos had served their purpose, inspiring Darwin to change direction and to re-examine all his specimens from the Galápagos until, finally, the pieces of the jigsaw were correctly assembled and his theory of natural selection was complete.

Furthermore, Darwin had been able to scotch one of the enduring myths concerning the giant tortoises. Not only was it now certain that the tortoises of the Indian Ocean and Galápagos were different from one another, but it was known that there were possibly several species within the Galápagos alone. However, tortoise speciation was no longer Darwin's problem. During the rest of his life there is little evidence that he ever gave much thought to the animals that inspired him to think of the idea with which he is now synonymous.

Ironies

W E CANNOT LEAVE the subject of the Galápagos tortoise and its place in evolution just yet because from Darwin's ten-year anguish on the subject comes a supreme irony.

It will be remembered that the seeds of Darwin's evolutionary ideas were planted when Governor Lawson bragged of being able to identify the island of origin of any tortoise placed before him. This is indeed a great boast because when modern herpetologists with years of experience are faced with a Galápagos tortoise, they are hard pushed to tell which island it came from without examining the animal in some detail. Indeed, when zoologists try to determine the origin of Galápagos tortoises in zoos around the world, they look at the animals' DNA rather than the shape of their shells. Lawson would certainly have had difficulty separating the tortoises into their island races merely by sight alone. What then provoked him to make such a claim? When we look at another Galápagos visitor's comments on the tortoises, we may well find the answer.

In his journal David Porter, a captain in the American Navy, describes two quite definite types of shell shape on the Galápagos tortoises – one like a 'Spanish saddle', the other 'round'. This matches very well with the modern description of the Galápagos tortoises, being either saddlebacked or domed in shape.

It is easy enough to tell a saddleback tortoise from a domed, but because both shell shapes belong to more than one species on different islands it is difficult to give the exact location of a tortoise solely based on this diagnosis. If modern experts cannot do it, it is unlikely that Lawson could either.

It seems much more likely that, like Porter, Governor Lawson was aware that the shells of tortoises could be either saddlebacked or

The shape of a tortoise's shell was important to naturalists. At the top is a dome-shaped tortoise from Isabela Island, while below is a saddleback from Pinta Island.

domed and that his claim was based on his observation that these two types were restricted to particular islands.

Darwin himself did not notice the different types of tortoise because the three islands that he visited were home only to dome-shaped tortoise species. Santa María Island, on which Darwin met Lawson, alone had tortoises with the classic saddleback shape but by 1835 this breed was practically extinct. All that Darwin saw of them were their empty shells, which had been upturned and made into flowerpots.

It is somewhat ironic that Darwin's belief that Governor Lawson could identify perhaps a dozen or more separate tortoise species has in fact no basis in reality. Indeed, so closely related are the various types of Galápagos tortoise that they have since been redefined as one species with fifteen separate subspecies (but more of that later).

Despite this irony, the scientific community must still give thanks to Governor Lawson, for if he and Darwin had not met on that desolate windswept island, and if he had not decided to impress his young guest with an idle boast, the evolutionary cause might have been set back by decades.

Before leaving the subject of Darwin's conversion altogether there is one other irony that is worth noting. Whereas he had too much faith in Governor Lawson's ability to distinguish types of tortoise, in the interval since Darwin made public his ideas on evolution, other scientists have put too much faith in his Galápagos work.

For many years the myth surrounding Darwin's time in the Galápagos led some scientists to believe that the young naturalist had coined the idea of natural selection whilst in the islands and had thus accurately recorded the location of his finch specimens on the spot. We now know that he did not and that he retrospectively labelled his specimens based on the information provided by Captain FitzRoy and others.

However, on at least three occasions curators at the British Museum have taken Darwin's (retrospective and sometimes inaccurate) labelling at face value and have used it to change the (genuine) island location data written on the tags of FitzRoy's specimens. Darwin was held in such high regard that it was felt that the locations of his birds must be accurate while those of FitzRoy must be in error. We now know differently. At least one ornithologist dismissed FitzRoy's location data in their entirety, believing them to be hopelessly inaccurate in comparison with Darwin's.

The labelling errors also led to some confusion about the real distribution of some bird species, as subsequent expeditions to the Galápagos had problems trying to match Darwin's species distribution to their own findings. This has led to some scientists pronouncing that certain species of finch had become extinct on particular islands simply because Darwin had erroneously recorded them as originating from there and they could not be found. There was even a belief that some new finch species had evolved in the decades between Darwin's visit to the Galápagos and those of subsequent scientific expeditions.

The mythology of Darwin and the Galápagos continues to fascinate and inspire biologists, geologists, zoologists and all manner of other scientists. It took over a century for fact to be separated from fiction and for that we must thank Professor Frank Sulloway, on whose work much of this chapter is based. It was Professor Sulloway who painstakingly raked all the available sources in order to understand what Darwin did and did not manage to achieve during his short time in the islands, and also how his Galápagos experience influenced his later thoughts and actions. Anyone who wants further information on Darwin's time in the Galápagos should begin by consulting Professor Sulloway's work.

PART THREE
Decimation

The Indian Ocean Tortoises

SIX MONTHS AFTER visiting the Galápagos, the *Beagle* pulled into Port Louis, the principal harbour of Mauritius and the island home of another race of giant tortoises. Darwin spent the following three weeks on Mauritius, during which time he covered much ground, collecting plants and animals from across the island. Despite this, not once does he mention having sighted a giant tortoise. This is not altogether surprising. If he had ever had any ambition of seeing a Mauritius tortoise, by the time of his visit he was already many years too late. By 1835 the wild populations of Mauritius tortoise had been extinct for many years.

In the sixteenth century there were seven islands in the Indian Ocean with native populations of giant tortoises. These were the Mascarene Islands (Mauritius, Réunion, Rodriguez), the Seychelles (Mahé, North Island and Astove Island) and Aldabra atoll. However, by the year 1800 only one island was home to any wild tortoises at all. Clearly the interceding years had seen a calamitous collapse in tortoise numbers. What could have caused such a grand disaster to occur in such a short space of time?

The paucity of written records makes reconstructing the history of the Indian Ocean tortoises extremely difficult but there is just enough information to allow us occasional glimpses as to what was happening to these animals and to provide a clue as to the cause of their demise. As was shown earlier, sailors passing the Mascarenes and the Seychelles often commented on the sweet taste of a cooked tortoise and also on their immense size. These attributes, whilst pleasing to the sailors, were unfortunate for the tortoises.

The Mascarene Islands, being on a key trade route from the southern Atlantic to India, were often the first port of call for ships travelling to and from the Cape of Good Hope. Naturally enough, when sailors had spent several weeks at sea eating nothing but fish, preserved meat and pickled vegetables, the tortoises were on the menu of every ship traversing the southern Indian Ocean and were eagerly sought after by ships' cooks and crews alike. The demand for the tortoises soon outstripped supply.

The voraciousness of the sailors' appetites can be illustrated by the fate of Mauritius' most famous extinct animal – the dodo. This strange-looking bird was actually a relative of the pigeon that, many thousands of years ago, became stranded on Mauritius. In the absence of any predators, the dodo's pigeon ancestors evolved to become tall, fat and flightless, which meant that, when the sailors arrived, they were a very visible food source and comparatively easy to catch.

This bird had not evolved to become one of nature's beauties, as is proved by the first written description of the dodo, by a Dutchman in 1601:

> Grey parrots are also common there, and other birds, besides a large kind, bigger than our swans, with large heads, half of which is cov-ered with skin like a hood. These birds want wings, in place of which are three or four blackish feathers. The tail consists of a few slender, curved feathers, of a grey colour. We called them Walg-vogels [dis-gusting birds], for this reason, that the longer they were boiled, the tougher and more uneatable they became.

Dodos were once abundant in Mauritius but if the giant tortoises were noted for their fine eating, then the reverse can be said of the dodo. Sir Thomas Herbert, an English visitor to Mauritius, once commented that the bird was 'reputed more for wonder than for food, greasie stom-achs may seeke after them, but to the delicate they are offensive and of no nourishment'. Despite its apparent unpalatable nature the dodo was mercilessly hunted by the Dutch, the Portuguese, the French and anyone else who set foot on Mauritius. Unbelievably, the bird was unafraid of man and was hunted by hand, being commonly strangled to death and then roasted on a spit. To add insult, the bird would often be basted in tortoise oil in an attempt to improve its flavour.

Logbooks from ships in the area refer to crews going ashore and gathering dozens of dodos at a time, pickling those that were not eaten straightaway. The dodo was not a prolific breeder, laying only a solitary egg, many of which were also gathered and eaten by the sailors. The rate of decline was rapid and the last confirmed eyewitness account of a living dodo was in 1662. It is assumed that they became extinct shortly afterwards.

In the case of the dodo it had taken the European sailors less than a century to eat their way through a species that was not only deemed inedible by many but which was once very common. The giant tortoises, which were also large, edible and easy to catch, were also sought out for food by sailors, suggesting that their extinction would not be all that far behind that of their feathered friends.

Many ships' accounts of dodo gathering mention that their crews went ashore to collect tortoises too, demonstrating that the fate of these two animals was not only entwined but also sealed fairly soon after the sailors' arrival. However, it was not in Mauritius alone that animals were being taken and eaten in large numbers. The same was true of the other Mascarene Islands where visiting ships would gather a hundred or more tortoises at a time.

In 1673, Hubert Hugo recounts that the ship on which he travelled took 400 to 500 tortoises from Mauritius, all of which were processed for their meat and fat. He also noted that pigs, which had been introduced into the island a few years previously, were eating the tortoises' eggs as well as the hatchlings. Even at this early stage Hugo comments that tortoises were becoming difficult to find in the wild.

The same process of devastation was enacted on Réunion and Rodriguez where the adult tortoises were slaughtered or taken as a live food source. Newly introduced omnivorous mammals, such as rats, cats, dogs, goats and pigs, were stopping the eggs from hatching or, if they did hatch, preventing the young from reaching breeding age. As for the dodo, it was a recipe for disaster.

By the mid-1700s all the Mascarene Islands had been colonised and the land was being cleared wholesale in order to make room for plantations and cattle farms. This land clearance was the final straw for the tortoises, which were being forced into those small and remote pockets of natural vegetation that remained. Reduced in number, cut off from one another and still being hunted, the tortoises were often unable to breed. Even if they could successfully mate, the eggs and hatchlings would be scavenged by pigs and dogs.

By the 1750s we have only a few reports of Mauritius and Réunion tortoises having been seen in the wild and it can be assumed that this poor creature was rapidly heading down the same path as the dodo. However, by this time the Mauritius colonists had built up an entire economy around the giant tortoise. Not only were they used to eating them but they had also made a small fortune selling them to passing ships. Faced with a lack of tortoises on their own island, the people of Mauritius began to look elsewhere for a fresh supply.

On the more remote and sparsely colonised Rodriguez Island, the giant tortoises had fared better than on Mauritius and Réunion, so much so that they were still relatively abundant in the 1750s. Any possibility that they might survive the slaughter taking place elsewhere was crushed when the people of Mauritius got wind of their presence and moved in for the kill. Between 1759 and 1761 alone, over 21,000 animals were captured and removed from Rodriguez to be sold for food in Mauritius. The trade continued for several years, exporting 4,000 to 5,000 tortoises annually. It does not take a genius to work out that such an industry was not sustainable for any length of time and by 1770 the tortoise trade was at an end, with too few animals to make it viable.

The shell of the extinct Rodriguez tortoise
(*G. vosmaeri*). The last report of a living
specimen was in 1795.

It is not clear exactly when the last of the Mascarene tortoises died out in the wild. In Mauritius the last report of a wild tortoise was made in 1778; on Rodriguez it was in 1795; while on Réunion it was as early as 1754. Although an exact date cannot be placed on the moment of extinction, it is generally agreed that by the turn of the nineteenth century there were no giant tortoises living in the wild on Mauritius, Réunion or Rodriguez.

Further north in the Indian Ocean, the Aldabra and the Seychelles tortoises fared somewhat better than those in the Mascarenes Islands, at least for a while. Being away from the main trade routes, these islands were less frequently visited and were not even discovered until several decades after Mauritius.

Aldabra is a barren atoll with only limited fresh-water supplies and it was far from being strategic. No one in their right mind would have tried to colonise it, and after its discovery in 1744 decades would pass before anyone bothered to set foot on it again. The Seychelles were a different matter. The main islands were mountainous and lush with tropical vegetation, ideal for plantations. Nonetheless, because of their remoteness, the first colony was not set up on the largest island, Mahé, until 1768. At this time the native giant tortoises were still noted to be abundant.

Unfortunately, the fate of the Seychelles tortoises and those in the Mascarenes would soon become tied to one another. As the Seychelles were being colonised, so the citizens of Mauritius were eating their way through the last of the Rodriguez tortoises and still their appetites were not sated. After the decimation on Rodriguez, Mauritius began to look elsewhere for tortoises.

The canny new colonists in the Seychelles, who found themselves surrounded by giant tortoises, spotted a gap in the market and so, in 1771, began to export their tortoises to Mauritius. Available government records show that around 10,000 animals were removed between 1773 and 1810 but this is almost certainly a radical underestimate of the true scale of the trade, as other ships were coming in and independently gathering their own supplies from remote locations.

In 1786 an official record from the Seychelles notes that on Ile aux Récifs there were once giant tortoises but 'private vessels have carried them off, so that but few remain', and that on Ile Praslin, 'land tortoises were formerly excessively common, until the crews of the trading vessels which called there for cargoes of these creatures, took to burning the scrub to find them.'

This trade in tortoises continued well into the 1830s and was probably prolonged by ships travelling to Aldabra to take tortoises from there to the Seychelles, from whence they would be exported to Mauritius. This mixing of tortoises from different islands makes it extremely difficult to pinpoint when the native Seychelles tortoise became extinct. In 1787 it was estimated that only 6,000 to 8,000 tortoises remained in the wild, which, at the rate of gathering and export then taking place, had a life expectation of only a few years.

It is probable that by the 1840s the native Seychelles tortoises had also vanished from the wild, leaving Aldabra as the last refuge of the Indian Ocean giant tortoises. It was conceivable that the local practice of keeping the animals as pets meant that some individuals might yet have survived in the gardens of a few wealthy Seychellois. However, as many of these 'pets' were only retained to form the main dish at wedding feasts, even their days were numbered.

Such was the quick pace of extinction that scientists were not able to examine any of these animals in the wild before their disappearance, and so exactly how many species of tortoise had been lost would not be worked out for many years. With the exception of the Aldabra tortoises, and possibly a few lone individuals being kept as pets, the human race had managed to eat its way through several islands' worth of giant tortoise in only a few short decades. It was not a record to be proud of. Nor was it to be the last time that these gentle giants were to find themselves the victims of human appetites. As the slaughter in the Indian Ocean was coming to its bloody finish, across the world another one was only just beginning.

The Safe Haven

B Y THE LATE eighteenth century, development and settlement had traumatised the giant tortoise populations on the Mascarenes and Seychelles but the same had not been true on the Galápagos Islands. This archipelago, stuck in the middle of nowhere, had been protected by its isolation and, unlike the Seychelles and Mauritius, for example, was well away from any busy shipping routes. The islands also had few natural resources to be plundered and the only visitors would be ships en route across the Pacific Ocean. With each one a few dozen tortoises would disappear into the sailors' stomachs but the effect on the overall tortoise population, which numbered many tens of thousands, was negligible.

In 1685, by which time the dodo was already extinct and many Indian Ocean tortoises were struggling for survival, the pirate William Dampier visited the Galápagos and recorded that as the tortoises 'exceed in sweetness, so do they in numbers; it is incredible to report how numerous they are'. The same was to be true for a further century. Then, without warning, the Galápagos Islands, hitherto ignored, suddenly became of vital importance to the world. The start of the problem was another variety of gigantic animal – the ocean-going whales – whose fortune in many ways parallels that of the tortoise. For a brief while, the fate of these two unlucky animals was to become entwined.

Like the tortoise, the whale could be exploited by man and by 1700 hundreds of different uses had been found for the various body parts of most species of cetacean. Their oil could be used to light lamps, their blubber to manufacture soap and their bones to support ladies' corsets. Such was the whale's usefulness that many governments became concerned that they might not be able to guarantee a steady supply (rather like the relationship of modern politics to crude oil). They therefore agreed to pay huge bonuses to whaling ships to encourage them to stay at sea longer and thus return with bigger catches. This, combined with the general high price paid for whale products, meant that there were fortunes to be made from the whaling industry. Throughout the eighteenth century whaling fleets flooded into the Atlantic Ocean from European and American ports, battling with one another for the favourite hunting grounds.

Although ships were prepared to travel as far as the Antarctic in search of whales, they always stayed in the Atlantic Ocean. With vessels from so many nations hunting so few whales, it was only a matter of time before a serious diplomatic incident occurred. In fact, it was to be a relatively small-scale confrontation but it started a chain of events that would eventually lead to disaster for the Galápagos Islands.

In 1789 the king of Spain became concerned about his country's dwindling whale catches and decided to clamp down on foreign fishing in Spain's South American coastal waters. Given that Spain (like other colonial powers) claimed territorial rights to waters stretching many hundreds of kilometres out from its coastline, this effectively excluded foreign whaling ships from practically the entire southern Atlantic. It also broke several pre-existing treaties, which allowed foreign vessels to operate along uninhabited coastlines.

The first victims of the new law were two British whaling boats that had anchored off Penguin Island in Patagonia. The Spanish Navy found them and quickly forced a surrender. Even though the British ships were not actively hunting and had only stopped for repairs, the Spanish confiscated both the vessels and their catches. This news was not greeted with joy. In parliament the Prime Minister summarised his government's feelings: 'If that claim were given to, it must deprive [Britain] of the means of extending its navigation and fishery in the southern ocean, and would go towards excluding his majesty's subjects from an infant trade, the future extension of which could not but be essentially beneficial to the commercial interests of Great Britain.'

The British readied themselves for war and started to assemble a task force but in the end it was not needed. At the thought of a costly naval war with Britain, the Spanish backed down, allowing UK vessels access not only to the waters on the Atlantic South American coastline, but on the Pacific side too.

Before this time the British had had no interest whatsoever in the Pacific Ocean. It was considered too distant, devoid of whales and too politically sensitive an area for shipping. However, with the Spanish now allowing the British into the Pacific, it would have been foolish for them not to at least find out whether there were any whales in those waters or not. In 1789 the whaling ship *Emilia* duly made her way around Cape Horn, with her sceptical captain fully expecting to reach her home port not having sighted a whale. He was wrong and only a few months later the *Emilia* returned, laden with a catch that included 140 tonnes of whale oil and nearly 900 seal skins. Furthermore, it turned out that the Pacific was teeming with sperm whales whose oil and other by-products were more valuable by far than those of any other whale. The *Emilia* had hit the jackpot and, after centuries of being left in peace, the Pacific sperm whales were now in the sights of the massive whaling fleets of Britain and America. However, the British still had one major difficulty.

Although at that time they had made a truce with the Spanish and North Americans, frequent and sudden hostilities would break out with these nations, so ceasefires were normally temporary. Between them, the Americans and Spanish owned the entire tropical Pacific coastline, which gave their boats a distinct advantage over the British, whose nearest home port was the recently founded penal colony in Botany Bay, Australia. Consequently the British whaling captains had no guaranteed safe haven to which they could run in the event of trouble. Therefore, while the Americans readied their whaling fleet, the British whaling companies hesitated. Even an increased government bounty on their catches could not persuade them into the Pacific.

In 1791, as American boats swarmed into the Pacific, the British government decided that it needed to act. Its solution was simple: if there was no current safe haven in the Pacific Ocean, they would just have to create one.

In 1792 the Admiralty approached a Captain James Colnett with a proposal to send him to the Pacific to seek out and claim new British territory. On paper Captain Colnett was an excellent choice.

He was one of the most experienced naval captains in Britain and had spent almost all his adult life at sea; he had even served with Captain Cook on his second circumnavigation of the globe. However, he also seemed to have a nose for trouble and had previously been captured by pirates and privateers as well as by the Mexican and Chinese navies. Colnett was getting on in years and was initially reluctant to take the commission, complaining that after his last voyage he had returned to England to find that 'during [the] interval, death had deprived me of my nearest relations'.

The Admiralty, knowing that, above all else, Colnett was interested in money, continually raised his proposed salary until, at last, the captain buckled and agreed to the venture. To help him with his task he was equipped with *Rattler*, a 374-tonne sloop of war, but before he had even left port Colnett had to deal with a minor mutiny among his crew. He then staged a mutiny of his own, refusing to leave until the Admiralty guaranteed him promotion upon his return. After much wrangling the *Rattler* departed for the south seas on 2 January 1793.

The ship made good time; by Easter the *Rattler* had made the perilous voyage around Cape Horn and was sailing in Pacific waters. The search for a British safe haven began. Colnett knew that any attempt to carve a British annexe on the Spanish-dominated mainland would ultimately result in a diplomatic incident, the end result of which would be the capture and destruction of the *Rattler*. So, instead of searching the coves and inlets of mainland South America, Colnett visited all the known offshore islands, the majority of which still had no fixed position on the charts. More importantly, they were still for the taking by whichever country could be bothered to claim them.

Colnett's first port of call was the exceedingly small and remote islands of Saint Felix and Saint Ambrose, located west of the tropical Chilean coast. Any hopes that he had of turning these into a naval base were quickly dashed. Both islands were steep, barren, waterless, almost completely devoid of shelter and surrounded by heavy surf. After a disastrous attempt at landing, during which a rowing boat capsized, killing a seaman, Colnett abandoned the islands. He was going to have to look elsewhere.

He headed north, hunting in vain for other islands that were sufficiently far from the coast to be classed as out of Spanish jurisdiction. Despite two months of searching, Colnett drew a blank. The task looked increasingly hopeless but, undeterred, he decided to fall back on his Plan B.

Like many seafarers of his generation, Colnett had avidly read the published journals of the renegade buccaneer William Dampier, who was more than familiar with South America and the Pacific Ocean. As we have seen, Dampier had also spent long periods hiding away in the Galápagos and had developed an attachment to these desolate islands. Colnett was exploring the Pacific a century after Dampier but the Galápagos were located in such an inconvenient place that they still remained uninhabited. More importantly, they were also uncharted and unclaimed by any nation. As Colnett leafed through Dampier's descriptions of his 'Gallappagos' islands, he stumbled upon his description of a natural harbour on San Salvador Island, which had 'excellent good, sweet water, wood, etc., and a rich mineral ore'. If this harbour was as promising as Dampier made it sound, perhaps the Galápagos could serve as the British safe haven within the Pacific. On 19 June 1793, Captain Colnett departed the South American coast and set sail for the Galápagos, arriving a few days later.

In common with almost every other sailor to have visited the islands before him, Colnett's first impressions of them were less than favourable. He saw no other ships there. Despite several days of searching, the *Rattler's* crew could find no food and just one pool of brackish water. In his journal Colnett complains bitterly about the lack of vegetables, fresh water and firewood. He believed that Dampier and his ilk had been lying when they gave descriptions of verdant bays and running springs. 'The Spaniards,' he wrote, 'are said to be acquainted with an island [which has] plenty of water on it . . . but I am not in the habit of giving an implicit faith to Spanish accounts.'

In addition to this, Colnett had not seen a sperm whale for days, which, to his mind, meant that the islands were too far from the traditional whaling routes to be of use as a bolthole during times of trouble. One week after their arrival, the crew of the *Rattler* set sail from the Galápagos to look elsewhere for their safe haven.

Despite the bad impression given by the Galápagos, the sailors made one major discovery: the giant tortoises. They had encountered them on the very first island where they landed but the thought of eating them seemed repellent. 'So disgusting is their appearance, that no one on board could be prevailed on to take them as food.'

A few days later, the threat of scurvy and lack of any other fresh food made Colnett and his men think again. The captain ordered that one of the beasts be captured, dragged on board, butchered and the meat then turned into soup. When the order had been carried out, it fell to Colnett to do the tasting, which he did with some reluctance. He need not have worried. Like many others before him, he discovered that the tortoises made a superb meal. In his journal he records that his tortoise had been turned into 'an excellent broth'. There may not have been fruit, vegetables, wood or water on the Galápagos, but at least there was one reliable source of food.

With his Galápagos visit deemed a failure, Colnett returned to the South American coast and continued northwards, hunting for uninhabited islands on the way. The result was a disaster. Cocos Island was too wet and full of rats, while the islands of Socorro, Sancta Berto and Roca Partida (now known as the Revillagigedo Islands) had been claimed by the Spanish. (Ironically, the Spanish had laid claim to the Revillagigedo Islands some years previously, after they had captured Colnett off the Mexican coast. This episode had prompted the Spanish to survey and assume proprietary control over all the islands along that particular coastline.)

In December, after reaching the Californian coast, Colnett gave up and began the long journey south again. His mission seemed increasingly likely to end in failure.

The route back down the coast was no less eventful. On Quibo Island, just off Panama, Colnett managed to get bitten by a snake and attacked by a giant alligator on successive days. As the alligator rushed at him, Colnett just had time to pull out his gun but the noise 'roused the hideous animal and he was in the act of springing at me when I discharged my piece at him, its contents entering beside his eye, and lodging in his brain, instantly killed him'. Despite an abundance of water, wood and food, that island was permanently struck off

Colnett's list. Not only were the animals dangerous, but the Spanish men-of-war were too close.

A few days out of Quibo, the *Rattler* encountered a large school of sperm whales and killed seven of them. Seeing so many whales in such a small geographical area made Colnett rethink his strategy. If the whales were to be found in the equatorial latitudes, this, logically, was where the British safe haven should be located, thus affording their whalers a great advantage. However, the only suitable place in that region that Colnett had found was the Galápagos Islands. It was with some misgivings that the *Rattler* abandoned her southern journey and set sail westwards towards the Galápagos once more.

On his arrival at the islands Colnett experienced a sense of déjà vu as he spent his first few days in a becalmed state, unable to find water, salt or substantial amounts of fruit and vegetables for the crew, many of whom were erupting in the boils of scurvy. Colnett was becoming increasingly disenchanted with the Galápagos and on 23 March, after yet another fruitless search for water, his patience snapped. 'The inhospitable appearance of this place,' he wrote in his journal, 'was such as I had never before seen nor had I ever beheld [except] in the fields of ice near the South Pole.' It was to be another couple of weeks before fortune smiled on him.

The first lucky break was the discovery that there were sperm whales in the vicinity of the Galápagos and in large numbers too. In fact, there were so many whales that Colnett believed that he had stumbled across the main breeding ground for all the sperm whales in the eastern Pacific Ocean.

With such a profusion of whales, it became obvious that the Galápagos Islands could serve the British very well indeed as a hunting ground. If he could only find a reliable source of water, they could also be the sought-after safe haven. Island after island proved to be bone dry and it was only after another two weeks that the trickle of a freshwater stream was found in a bay on San Salvador Island. Near to the stream was evidence of an abandoned encampment, probably made earlier in the century by buccaneers. Evidently whoever had built it had managed to survive for some time on the stream's water, indicating that the spring that fed it was probably permanent.

This was enough for Colnett. The discovery of fresh water on the Galápagos meant that, as far as he was concerned, his quest for a safe haven was over. The Galápagos Islands were no paradise but they had an abundance of whales to hunt, some wood to burn and a limited amount of water.

'These isles,' he wrote, 'deserve the attention of the British navigators beyond any unsettled situation but the preference must go to James [San Salvador] Isle as it is the only one [where] we found sufficient fresh water to supply a small ship.'

When it came to food supplies, Colnett commented that the ocean was full of fish but that

> all the luxuries of the sea yielded to that which the island afforded us. This is the land tortoise which, in whatever way it was dressed, was considered by all of us as the most delicious food we had ever tasted . . . Vessels bound round Cape Horn to any part north of the Equator, or whalers on their voyage to the north or south Pacific Ocean, or the Gulf of Panama, will find these islands very convenient places for refitting and refreshment. They would also in future serve as a place of rendezvous for British fishing ships, as they are contiguous to the best fishing grounds.

Convinced that his quest was over, Colnett set sail from the Galápagos on 13 May and headed south on the long voyage home to England. In November the *Rattler* arrived back in the United Kingdom after an absence of twenty-two months. The Admiralty were anxious to hear Colnett's report. Where were the British whaling ships to go? Colnett had but one answer. They should go to the Galápagos, he said.

The Arrival of the Whaling Fleet

I N 1795 THE first British whaling ships rounded Cape Horn and entered the Pacific Ocean. Thanks to James Colnett, the captains knew that they should head directly to the Galápagos Islands where they would find shelter, sperm whales, water and food.

The Galápagos were still uninhabited and rarely visited when the first British whaling boats began to anchor in the natural harbours there. Especially popular was 'Port Rendezvous', a sheltered cove on the western side of Isabela Island, which had deep water that ran almost up to the beach itself, and 'Post Office Bay' on Santa María Island where a barrel nailed to a tree served as a postbox. This postbox, which was there in Colnett's time, was an informal arrangement whereby ships that had just entered the Pacific would leave letters handed to sailors by the relatives of whalers already in the region, while homeward-bound ships would collect and deliver any letters for people back at home.

As the popularity of the Galápagos grew, with both British and, later, American ships, so it would be common to find two or three vessels at anchor in either of these locations. After centuries of isolation, these remote islands had come to the attention of the world at large and were starting to play a part in a valuable economic trade. From this point onwards the Galápagos would never again be totally deserted. Even though it would be decades before a permanent colony was established, there would always be several ships in the islands' waters at any one time. Although the Galápagos were a convenient place to drop anchor for a while, most of the ships came for one reason only: to collect tortoises.

Since the animals' accidental discovery in 1535, the Galápagos tortoise population had barely been touched by the activities of visiting humans. Unlike the Indian Ocean tortoises, in 300 years the bucca-neers and military ships had only taken a few dozen tortoises at any one time, and their visits were so infrequent that the effect on the tens of thousands of animals inhabiting the islands was negligible. Thus, in the 1790s the first whaling ships to arrive at the Galápagos were entering a near-pristine ecosystem. Seventy years later, when the last whalers left, that ecosystem was near to collapse.

One of the real dangers of a long sea voyage was the risk of scurvy, a terrible disease that progressively weakens and destroys the body. Although ships setting out from their home ports could take with them preserved vegetable and meat supplies, these could not be expected to last more than a few months. As some of the whaling expeditions could be away at sea for two or three years at a stretch, after the preserved food ran out, supplies would need to be restocked regularly. Even if ships could find food, it was often not possible to preserve it (despite the marine environment, large quantities of salt are hard to come by), which meant that fresh food needed to be found every two or three weeks. In an area as vast as the Pacific this was simply not practicable and inevitably crews would become mal-nourished. At a stroke the tortoises solved this problem.

It was unfortunate enough that the giant tortoises should be so pleasing to the human palate but the whalers discovered another remarkable property of these magnificent animals: they could be kept alive on board for months and still be edible at the end of that time. If there was ever one factor that sealed the tortoises' fate, this was it.

'I have,' wrote one ship's captain, 'had these animals on board my own vessels from five to six months without their once taking food or water [and] they have been known to live on board some of our whale ships for fourteen months.'

Another captain recalled: 'Shortly after the ship *Niger* of New Bedford left the Galápagos, one of the tortoises disappeared. Two years later when the ship arrived at New Bedford, the tortoise was found alive among the casks in the lower hold.'

The tortoises were a captain's dream come true. Not only did they taste delicious, which meant no complaints from the crew, but also they could be stored alive for several months at a time without need-ing to be fed or watered. With a good supply of tortoises on board,

A Galápagos tortoise wanders the deck of a ship. Many tortoises spent the
last months of their lives waiting for the cooking pot.

a ship could spend weeks at a time away from land, hunting whales
in the remotest parts of the ocean. It was not long before news of this
miraculous food source spread among the British and American
whaling crews, causing dozens of ships to arrive at the Galápagos
with the sole intention of stocking up with tortoises.

The sailors referred to the tortoises as 'terrapins', the pronuncia-
tion of which quickly became distorted into 'turpin' or 'terapen'.
Soon the whaling ships' logs were full of references to the crew going
ashore to indulge in some 'turpining'. An entire tortoise-gathering
subculture grew up among the Pacific whalers.

A typical 'turpining' session would take place across several days,
beginning with the whaling ship anchoring just offshore. On the call
of 'all hands a-turpining', the rowing boats would be lowered, carrying as many crew as they would hold with instructions to go in
groups and find as many tortoises as possible.

For the first few years the whaling ships would have been able to
gather tortoises with relatively little effort as there were so many of
them milling around the lowland areas just behind the beaches.
However, the majority of them lived much higher up in the mountains where the air was cooler and the vegetation thicker. It did not

take long for sailors to eradicate the lowland tortoises, forcing the men to make lengthy and arduous forays into the islands' interiors. However, undertaking long journeys away from the boat presented a serious logistical problem of its own as the tortoises were so large that it was difficult to transport them from the highlands back to the ship. One solution would have been to kill them and strip them of their meat, leaving behind the heavy shell and bones, but this negated the true value of the 'turpin', which was the fact that it could be stored alive for months. Therefore, an immense amount of hard work was needed in order to carry or drag the tortoises down to the seashore and to load them on to the boats. The effort of moving the animals meant that the largest individuals were often spared by the collectors as they were simply too heavy to shift.

A full account of a 'turpining' expedition is provided by William Davis, captain of the American whaler *Chelsea*, who visited the Galápagos a number of times during the 1820s. On his first voyage there, Captain Davis anchored his ship off Santa María Island and went ashore with his crew. After leaving two men to prepare a camp, Captain Davis and the rest of his crew moved into the interior to find tortoises. After an unspecified length of time walking, Davis and his men stumbled on their prey.

'Presently to my surprise,' wrote Davis,

I saw our [Negro] Zekiel sitting on the rear of an enormous terrapin the size of a wheelbarrow, and much the shape of my mother's forty-gallon apple-butter kettle. Here was a beast that would weigh three hundred pounds at least. In the vicinity were numbers of others of more manageable size, and we selected two perhaps fifty pounds [twenty-two kilograms] in weight. We tied the fore and hind legs of each so as to leave convenient loops through which to slip our arms, intending thus to carry our capture home, knapsack-fashion, on our backs. [However,] the true way to carry a terrapin is to form a hand-barrow with deal clubs or, for the largest, of the steering oars, such a contrivance, manned by two or ten men, will bring down the capture with comparative ease.

[In the interior of the island] great terrapins were about, some of them of immense size – very much larger than any seen on the shore plains here. We took the head off the largest terrapin we could find;

one great enough to furnish a feast for a hundred men. We were exceedingly thirsty, moreover, and had tried to satisfy our craving with the warm insipid juice obtained from the trunks of the giant cactuses, but in our capture, in our terrapin, we found the living spring of the wilderness. An ample supply of pure limpid water was discovered in the pearly sack placed at the base of the animal's neck. There were some three gallons [thirteen litres] of water here and, wonder of wonders, it was cool . . . With one hundred and fifteen terrapin of all sizes secured, we then returned to the ship whose decks were covered with our sleeping captives and the cook's galley steamed with a new and savoury odour.

The few detailed descriptions of 'turpining' expeditions that have come down to us are all similar to this. Elation at finding tortoises soon gives way to weariness and thirst as the beasts have somehow to be carried to the ship, but there were other dangers too.

In 1841 the logbook of the whaler *Chili* recorded that a man had become lost during such a mission. Despite two days of searching, the lost sailor could not be found. Finally, on the third day, the *Chili* departed, leaving their man behind. 'Left bread and water,' wrote the captain, 'and directions in a bottle if anyone should ever find him.'

Each whaling boat would load as many tortoises as their decks and holds could handle. In most cases this would be between about twenty and fifty individuals but catches of between 100 and 300 – even, in one case, 500 – animals were not unheard of. Given that there could be several whaling ships at a time taking this number of tortoises, it must have been obvious that their numbers were going to decline rapidly, and so they did.

The first populations to suffer were those on the smaller, less mountainous islands, and also those tortoises living on the lowland areas of all the islands. Although exact figures are not available, it is clear that by the early 1830s ships were already having difficulty in obtaining sizeable numbers of tortoises from these areas. Even though the whaling ships had only been visiting the Galápagos for a couple of decades, the tortoise populations were already showing signs of distress, with catches on Santa María and Española islands decreasing from hundreds of animals in the early 1800s to a few tens by the mid-1830s.

Estimates of how many tortoises were taken during the early days of whaling are hard to come by but Charles Townsend, who studied many early logbooks from American whalers, reckoned that around 100,000 tortoises were removed from the Galápagos before 1830.

The heyday of the Pacific whaling fleet was between 1795 and 1815. During this time there were over 700 American whalers working the world's oceans, many of which were hunting the Pacific Ocean sperm whales. The other great Pacific whaling fleet, that of the British, was finding it difficult to operate so far from home. Even with the Galápagos as a safe haven, the long periods at sea took their toll on the crew and few volunteers were prepared to go to the Pacific. 'In 1801,' wrote one whaler, 'the Pacific sailors were the best in the world. By 1844 they were of the worst description.' Faced with such problems, the British interest in the region began to decline, and from around 1804 onwards there were progressively fewer British whale-ships.

~

In the modern world the British and Americans are firm allies but this was not always the case. In the first few decades of independence, the United States was at loggerheads with the former colonial power.

One particularly serious spat occurred in 1812, leading to a declaration of war between America and Britain. The Galápagos Islands were also caught up in this conflict, thanks largely to the efforts of Captain David Porter, whose maverick behaviour was to have an important bearing on the fate of the tortoises.

Captain Porter departed the east coast of America in October 1812 in the navy frigate *Essex* with instructions to rendezvous with Commodore Bainbridge in the South Atlantic. However, Porter disobeyed his orders and instead rounded Cape Horn with the sole intention of routing the British Pacific whaling fleet. Naturally, the first place that he headed for was the Galápagos Islands, the unofficial base for British ships in the Pacific.

Porter was remarkably successful and used a number of cunning tricks to lure the British whalers close to the *Essex* so that he could capture them without bloodshed. As well as flying the British flag and

David Porter, the maverick US Navy captain whose keen eye
provided the outside world with its first detailed description
of a giant tortoise.

continually disguising the *Essex*'s appearance, Porter also stole letters
from the barrel at Post Office Bay so that he could learn the number
of ships in the area and what their plans were. At the same time,
Porter would leave behind false letters suggesting rendezvouses with
British whalers, which, if the ships kept them, would result in their
seizure.

By these means Porter single-handedly managed to capture dozens
of British ships (as well as the occasional Spanish ship), and in the
process collected over $2.5 million of cargo. The *Essex* had a crew of

only 250 and Porter rapidly ran into difficulties when it came to man-ning the captured vessels. At one point he was so short-staffed that he placed David Faggagut, a twelve-year-old midshipman, in charge of one of the British ships. (Porter's choice was inspired, as years later Faggagut was to become the first full admiral in the US Navy.)

Much of Porter's time was spent in or around the Galápagos Islands but, unlike most sailors in the region, the captain became ena-moured of them, especially their wildlife. His journals contain some of the finest observations on Galápagos fauna to have been written up to that date and at the centre of his attention were what he called 'the elephant tortoises'.

We have already noted that Captain Porter was the first person ever to realise that the tortoises on different islands were markedly different from one another, a characteristic that Charles Darwin used to good effect in formulating his early theories on the Galápagos tor-toises' speciation, but the hardened sea dog seems to have felt a gen-uine affection for the shelled reptiles. In his writings Porter devotes many pages to describing the tortoises and, despite his continual dis-paraging remarks about their lack of beauty, underneath is a real fas-cination with their appearance and behaviour:

'Nothing,' recorded Porter in his journal,

perhaps, can be more disagreeable or clumsy than they are in their external appearance. Their motion resembles strongly that of the ele-phant; their steps slow, regular and heavy; they carry their body about a foot from the ground, and their legs and feet bear no slight resem-blance to the animal to which I have likened them; their neck is from eighteen inches to two feet [forty-five to sixty centimetres] in length and very slender – their head is proportioned to it, and strongly resem-bles that of a serpent. But, hideous and disgusting as is their appear-ance, no animal can possibly afford a more wholesome, luscious and delicate food than they do. The finest green turtle is no more to be compared to them in point of excellence, than the coarsest beef is to the finest veal; and after once tasting the Galápagos tortoises, every other animal food fell greatly in our estimation. These animals are so fat as to require neither butter nor lard to cook them, and this fat does not possess that cloying quality, common to that of most other ani-mals. When tried out, it furnishes an oil superior in taste to that of the olive. The meat of this animal is the easiest of digestion, and a quantity of it, exceeding that of any other food, can be eaten without experiencing the slightest inconvenience.

This passage reveals that, while Porter had an interest in the tortoises' natural history, he and his crew also had intentions of eating them. Porter provides us with a lucid description of the sort of damage that a large ship could do the Galápagos tortoise population in only a few short hours:

> The [turpining] parties (which were selected every day, to give all an opportunity of going on shore), indulged themselves in the most ample manner on tortoise meat (which for them was called Galápagos mutton), yet their relish for this food did not seem in the least abated, nor their exertions to get them on board in the least relaxed, for everyone appeared desirous of securing as large a stock of this provision as possible for the cruise.
>
> Four boats were dispatched every morning for this purpose, and returned at night, bringing with them twenty to thirty each, averaging sixty pounds [twenty-seven kilograms]. In four days we had as many on board as would weigh about fourteen tons [fourteen tonnes], which was as much as we could conveniently stow. They were piled up on the quarter-deck for a few days, with an awning spread over to shield them from the sun, which renders them very restless, in order that they might have time to discharge the contents of their stomachs; after which they were stowed away below, as you would stow any other provisions, and used as occasion required. No description of stock is so convenient for ships to take to sea as the tortoises of these islands. They require no provisions of water for a year, nor is any farther attention to them necessary, than that their shells should be preserved unbroken.

As well as giving us an idea of the numbers being taken, Porter's meticulous notes also give us a clue as to the real damage that was being inflicted on the tortoises' population structure. After one 'turpining' expedition to San Salvador Island, Porter reported:

> The most of those we took on board were found near a bay on the northeast part of the Island, about eighteen miles from the ship. Among the whole only three were male, which may be easily known by their great size, and from the length of their tails, which are much longer than those of the female. As the females were found in low

sandy bottoms, and all without exception were full of eggs, of which generally ten to fourteen were hard, it is presumable that they came down from the mountains for the express purpose of laying. This opinion seems strengthened by the circumstance of there being no male tortoises among them, the few we found having been taken a considerable distance up the mountains.

Naturally, the whalers and other visiting sailors were collecting those tortoises that were the easiest to catch, which, given the mountainous geography of the Galápagos, meant those in the lowland areas next to the beaches. However, what had gone unnoticed by all save Captain Porter, who had taught himself how to sex a giant tortoise, was that almost all the lowland tortoises were females. As Porter rightly surmised, this is because the females would wander down from the highlands in order to find suitable nesting areas in which their eggs could be laid.

This behaviour, which had served the tortoises so well across thousands of years, was now disastrous as it meant that the ships were preferentially harvesting the females from the lowland areas, leaving the males stranded up in the highlands. At the high rate of gathering taking place in the early 1800s, it did not take long before the female tortoises became extremely scarce in the wild, crippling their ability to breed and thus further limiting any chance that they might have had of recovering their numbers once the whaling ships disappeared. Indeed, a century later, when the first rough censuses of tortoise numbers were undertaken, most of the animals were found to be male.

None of these factors worried Captain Porter too much. Although his journals contained the most detailed descriptions of the Galápagos and its wildlife to date, he rarely reached any conclusions about what he had seen. After discovering that the tortoises had cold blood, he wrote that he would 'leave it to those better acquainted with natural history to investigate the cause. My business is to state facts, not to reason from them.' In truth, however, by the time he left the Galápagos there were probably few (if any) people on earth who were as knowledgeable about the islands' plants and animals as Captain Porter.

There were also probably few people who had eaten their way through as many Galápagos tortoises as Captain Porter and his crew.

According to his records, the *Essex* collected tortoises from Española, Isabela, San Salvador, Santa María and Santa Cruz, and probably killed in excess of 2,000 animals. They even stole them from the ships that they had captured and, on one occasion, fished fifty or so out of the sea after they had just been jettisoned by a fleeing British whaler.

Despite his capacity for eating them, the warrior-like Porter did the tortoises a minor favour. His constant harassing of the British cost them over 400 sailors and an estimated $2,500,000 in lost trade. In the end he was directly responsible for the whaling fleet withdrawing from the Pacific region. However, in taking on the British, Porter had stirred up a hornets' nest and he soon found himself being pursued by the *Phoebe*, a British frigate that heavily outgunned the *Essex*.

Porter played a game of cat and mouse about the Pacific but a heavy storm forced him to put the *Essex* into the Marquesas Islands for repairs. Here he was corned by the *Phoebe*, which gave Porter an opportunity to surrender by displaying the message (in flags): 'God and country, British sailors' best rights: traitors offend both.' Porter, still in defiant mood, replied: 'God, our country, and liberty; tyrants offend them.' Shortly afterwards the *Phoebe* opened fire, severely damaging the *Essex* and injuring or killing many of her crew. Porter, one of only two *Essex* officers to emerge from the battle intact, was taken prisoner but the next day stole a boat and made good his escape, leaving behind the message: 'Most British officers were not only destitute of honour, but regardless of the honour of each other.'

Despite being pursued, Porter managed to make his way back to New York with expectations of being treated like a hero by the navy. Unfortunately he found himself in receipt of a severe reprimand for having disobeyed his original orders and taken the *Essex* into the Pacific in the first place. Nevertheless, he managed to keep his rank and was soon harrying the Spanish until, in a complete volte-face, he left the American fleet and became head of the Mexican Navy.

Aside from his precise observations, Porter also routed almost the entire British whaling fleet and thus terminated a period of sustained pressure on the tortoises' numbers. The American whaling ships still kept coming but the discovery of whales off Japan shifted the fleet's location away from the Galápagos so that their presence decreased significantly until 1819, offering some respite to the tortoises.

According to Charles Townsend's figures, the number of animals taken by whaling ships between 1830 and 1900 was around 13,000, in comparison to his estimate of 100,000 taken in the few decades prior to 1830 (both figures are probably underestimates).

Even though the easing of the tortoise plunder helped to preserve their numbers, much of the real damage had already been done. On some islands almost all the female tortoises had been taken, leaving an unbalanced population with no means of quickly replenishing itself. On other islands the numbers plundered were so great that there was little hope of a breeding pair being able to find each other in order to mate in the first place.

The progressive scarcity of tortoises in general is recorded by the decreasing numbers harvested by the whaling boats and by the increased length of time it took the 'turpining' crews to collect a useful number of the animals. For example, in 1831 the crews were collecting 200 to 300 tortoises in five or six days. By 1860 in over a week they had gathered fewer than a hundred, with some ships obtaining only sixty or so. By this stage, crews were having to travel many kilometres into mountainous territory; alternatively they would have to buy tortoises from other crews or from the scattered settlers on the islands. On some of the smaller islands the tortoises were so few and far between that the effort needed to cull them was no longer worth it and all collecting was abandoned.

We should not forget the plight of the sperm whales, which drew the ships to the Galápagos in the first place. Their numbers too suffered dreadfully but as they were swifter and freer to move about than the tortoises, significant numbers managed to avoid the whalers, allowing the species to recover once the slaughter ceased.

The event that was to finally remove the last few whalers from the Galápagos occurred on 27 August 1859 when Edwin L. Drake struck oil in Crawford County, Pennsylvania. At a stroke the world largely abandoned whale oil as an energy source in favour of the more abundant and cheaply extractable crude oil. The days of heavy whaling were over and, as the international oil industry rose in stature, the American fleet decreased to almost nothing, making whale-ships a rare sight around the Galápagos Islands.

This was, of course, good news for the tortoises but by then they already faced another problem. All the attention paid to the islands by the whalers and by naval expeditions such as Captain Porter's had finally come to the notice of a few brave souls who were prepared to risk making a living on the islands. The tortoises were no longer alone – there were now humans living permanently on the Galápagos.

Settlers

THE SEYCHELLES AND Mascarene Islands had much to recommend them to settlers: tropical beauty, lush verdant landscapes and a position on strategic trade routes, so by the 1700s they had been permanently colonised. As part of that process the land had been cleared for agriculture and the native wildlife killed for food or sport. The influx of settlers spelt doom for many species, including the giant tortoises. By 1800 there were only a few straggling individuals left from the many races that had once existed on these islands. Only on the atoll of Aldabra, which was too remote and barren to be effectively settled, was a sizeable population of tortoises left in the wild. It was apparent that humans and tortoises could not live together in perfect harmony.

The dawn of the nineteenth century saw the giant tortoise populations on the Galápagos Islands still relatively untouched. The whaling boats had yet to arrive in any number and, more crucially, the islands had yet to receive any permanent settlers. For as long as this was the case, the tortoises could comfortably survive, especially those living in the remote highland interiors. In an ever-shrinking world it was inevitable that sooner or later a colony would be set up on these desolate islands.

The first recorded inhabitant was a lone Irishman named Patrick Watkins who around 1807 was ejected from an English ship on to Santa María Island where he was forced to take up permanent residence. Watkins was evidently quite a character as this description, provided by Captain Porter, reveals in lurid detail:

> The appearance of this man, from the accounts I have received of him, was the most dreadful that can be imagined; ragged clothes, scarce sufficient to cover his nakedness, and covered with vermin; his red hair

and beard matted, his skin much burnt, from constant exposure to the sun, and so wild and savage in his manner and appearance, that he struck every one with horror. For several years this wretched being lived by himself on this desolate spot, without any apparent desire than that of procuring rum in sufficient quantities to keep himself intoxicated, and, at such times, after an absence from his hut of several days, he would be found in a state of perfect insensibility, rolling among the rocks of the mountains. He appeared to be reduced to the lowest grade of which human nature is capable, and seemed to have no desire beyond the tortoises and other animals of the island, except that of getting drunk.

Watkins had a dark side to his character and was known to have tricked ships into sending crews ashore to collect potatoes and pumpkins from his vegetable plot, only to destroy their landing boats and rob the men. On one occasion he got a landing party so drunk that he hid four unconscious crew members and let their ship sail away without them. These men remained on Santa María Island as Watkins' slaves until, in 1809, he tricked another passing ship into visiting his vegetable plot and, in the crew's absence, stole one of their landing boats. He and his four 'slaves' set off from the Galápagos, leaving behind a farewell note:

> Sir,
> I have made repeated applications to captains of vessels to sell me a boat, or to take me from this place, but in every instance met with a refusal. An opportunity presented itself to possess myself of one, and I took advantage of it. I have been a long time endeavouring, by hard labour and suffering, to accumulate wherewith to make myself comfortable, but at different times have been robbed and maltreated, and in a late instance by Captain Paddock, whose conduct in punishing me, and robbing me of about 500 dollars, in cash and other articles, neither agrees with the principles he professes nor is it such as his sleek coat would lead one to expect.
> On the 29th May, 1809, I sail from the enchanted island in the Black Prince, bound to the Marquesas. Do not kill the old hen; she is now sitting, and will soon have chickens.
> FATHERLESS OBERLUS [This was apparently one of Watkins' pseudonyms.]

Watkins later turned up in the Ecuadorian port of Guayaquil, claiming that his four crewmen had perished during the voyage. It was, naturally enough, assumed that Watkins had killed them in order to save the fresh water on board, so he was immediately locked up in the local jail and was still there several years later.

After the departure of the eccentric and dangerous Watkins, the islands would again be deserted for another two decades until 1830. In that year, and after much bloodshed, Ecuador broke away from the newly independent Spanish province of Grand Colombia and declared itself to be an independent country.

One of the heroes of the battle for Ecuador's independence was José Villamil, a Louisiana trader who became embroiled in the Ecuadorian battles after settling there in 1810. On several occasions during the various wars with the Spanish, Villamil distinguished himself, earning the respect and admiration of the Ecuadorian political masters.

Early in the war the Galápagos had come to Villamil's attention after a number of enemy boats had fled from the Ecuadorian coast in order to seek shelter there. Villamil might therefore have thought that the Galápagos could be used as a base by rebel boats and should consequently be brought under Ecuadorian control. Whatever the reason, from the moment that Ecuador became independent Villamil pressed Juan José Flores, the first Ecuadorian president, to send a flotilla to the Galápagos Islands to claim them.

The President looked favourably on the plan and in January 1832 the schooner *Mercedes* sailed from Ecuador under the command of Colonel Ignacio Hernández. On 12 February Hernández claimed the Galápagos for the Republic of Ecuador. Although the British had had a nominal interest in the islands because of their use to the whaling industry, the loss did not trouble them. British ships would still be able to call there in order to take on water and tortoises.

Villamil was perfectly aware that Ecuador's claim on the islands would be difficult to enforce unless the country had a permanent presence there. With this in mind Colonel Hernández remained behind on Santa María Island, together with a group of rebellious soldiers who had been been spared the death sentence on condition that they went to the Galápagos as colonists. In time many said that had they known how desolate the islands were, they would have opted for death. The tortoises, on the other hand, had no say in the matter. The arrival of the colonists was to be the single biggest disaster to befall them since the arrival of the whaling ships in the 1790s.

While the *Mercedes* headed back to Ecuador, Hernández and his soldiers scouted Santa María Island. High in the mountains, they chanced upon a wide, flat plateau with a reliable source of water near by. At 300 metres, the plateau was cooler than the coastal plain and its soil was deep and fertile. Hernández instantly knew that he had found a suitable location for Villamil's colony. He named the place Asilo de la Paz or 'Haven of Peace'.

Two further boatloads of settlers (again, mostly political undesirables) arrived in April and June, followed by another in October of eighty colonists, headed by Villamil itself. This not only put the Santa María Island colony on the map but also, for the first time, introduced domestic animals on to the Galápagos. It was this action, above all others, that was most to injure the unfortunate tortoises. Donkeys, goats, pigs, cattle, cats and dogs were all offloaded on to Santa María Island with the colonists. Villamil himself stayed on as governor of the Galápagos and under his control the new colony prospered by trading tortoises, fruit and vegetables with passing whaling ships.

By the time of Charles Darwin's arrival in 1835, Santa María Island had a population of about 300 men and women, and was doing well for itself. In the absence of Villamil, the Englishman Nicholas Lawson was in charge and, as was explored in detail earlier, it was a chance comment of his that started Darwin down the road to his theory of natural selection.

Right from the outset the colonists had become highly dependent on the tortoises as a source of food and as a tradable commodity. In the three years between the colony's establishment and Darwin's visit, the colonists had managed to eat their way through a sizeable portion of the tortoise population on Santa María Island. 'The inhabitants here,' wrote Darwin in his diary,

> lead a sort of Robinson Crusoe life. The houses are very simple, built of poles and thatched with grass. Part of their time is employed in hunting wild pigs and goats with which the woods abound. From the

climate, agriculture requires but a small portion. The main article, however, of animal food is the terrapin or tortoise. Such numbers yet remain that it is calculated two days' hunting will find food for the other five in the week.

Despite this optimism the tortoises on Santa María Island were already becoming scarce. The governor therefore ordered the colonists to go to San Salvador and Santa Cruz islands to obtain more to ensure that there would be plenty to sell to passing ships. Catching tortoises on the other islands involved setting up camp for days at a time and so, wherever the colonists went, their domestic animals went too. Inevitably some escaped, spreading the nuisance of feral animals further afield.

While Villamil was trying to establish a viable settlement, the Ecuadorian government was busy using the islands as a penal colony. All manner of political exiles and lawbreakers were shipped out to Santa María Island or to the newly established colony on San Cristóbal Island.

The tortoises were now under threat from all directions. The whaling ships had collected those individuals that were easy to reach; now the colonists were scouring the more inaccessible parts of the islands for them. Meanwhile, the colonists and the whalers had successfully introduced non-native domesticated animals on to all the main islands.

Many of these introductions were accidental but the consequences were nonetheless catastrophic on the local environment. In 1813 the unstoppable Captain Porter was himself guilty of introducing goats on to San Salvador Island when, after landing, he failed to tether three females and one male. 'Future navigators,' mused Porter,

> may perhaps obtain here an abundant supply of goats' meat, for unmolested as they will be in the interior of this island, it is probable their increase will be very rapid. Perhaps nature, whose ways are mysterious, has embraced this first opportunity of stocking this island with a race of animals who are, from their nature, almost as well enabled to withstand the want of water as the tortoises with which it now abounds.

Porter was perfectly correct. The goats multiplied at a prestigious rate, as they did on every island on which they had been landed, and started eating their way through much of the tortoises' vegetative food supply. The same was true of the pigs and cattle, while the carnivorous and omnivorous animals, such as the cats, rats and dogs, ate the tortoises' eggs and the delicate young hatchlings. A full-scale ecological disaster was in progress. The tortoise numbers were in severe decline.

Around 1840 the Santa María Island tortoise became extinct. Similarly, during the 1840s, San Salvador and Santa Fe islands contained so few tortoises that the colonists no longer bothered going there to collect them. A glimpse of the scale of this slaughter can be seen in 1847 when one settler on San Cristóbal Island claimed to have a personal stock of 300 tortoises available for sale to passing ships.

A small respite came in 1837 when the governorship passed from Villamil to Colonel José Williams. The new governor turned out to be something of a dictator, forcing the permanent settlers to sell their vegetables and tortoises to him at a much reduced price so that he could sell them on to visiting ships, creaming off the profit for himself. Faced with this tyranny, the free settlers began to desert the islands for the mainland but the deported prisoners had no such choice. Protests to the Ecuadorian government drew no response. Eventually the pressure grew until, on 6 May 1841, a revolt occurred, and the despotic Williams and his mercenaries fled the islands. At this point there were only fifty-eight colonists left and by the 1850s this had dropped to only twenty-five or so.

The colonies may have then dwindled to nothing but the real damage had already been done. In only a couple of decades an island race of tortoise had been rendered extinct and feral mammals now existed on all the main Galápagos Islands. Even without the settlers these multiplied at a prodigious rate, wreaking havoc on the environment. Villamil's colonial ambitions had opened a Pandora's box in the Galápagos and it would take another century before it could be partially closed again.

During the 1850s the Galápagos underwent another period of political instability. Ecuador, faced with mounting debts to Great Britain, offered the islands in lieu of payment but the plan was abandoned after protests from the French and Spanish. A similar plan to lease the islands to the USA was vetoed after protests from Great

Britain, Peru, France and Spain. The issue of piracy also raised its head after the privateer Birones was deported to Santa María Island but he and his crew managed to steal a whaling ship and return to Ecuador, killing one of the senior generals there before being captured and hanged.

The first truly successful settlement of the Galápagos did not occur until 1869 when a few hardy Ecuadorians arrived in order to start a plantation on San Cristóbal and to gather the commercially valuable moss. The plantation was successful and, after a new penal colony on Santa María Island had collapsed in a bloodbath, became the principal settlement in the islands. Over the coming years more settlers took a risk and moved out to the Galápagos. By the turn of the twentieth century there were permanent colonies on the islands of San Cristóbal, Santa María and Isabela.

The arrival of waves of new settlers meant more disruption for the tortoises. Land was cleared for agriculture and grazing, reducing their habitat; the adult tortoises themselves were still being harvested for their oil and as food, and their young were prey to feral animals.

By the 1870s both the Aldabra and Galápagos tortoises were struggling to survive but their plight was going totally unnoticed in the civilised world. Few people had even heard of Aldabra or the Galápagos, let alone their tortoises, and the scientific community was still entirely ignorant on the subject of how many species there were, where they lived or how many were left in the wild. The world's giant tortoises urgently needed help before they disappeared altogether. Fortunately they were to receive a champion, if a somewhat unconventional one.

PART FOUR

Obsession

La Mare aux Songes

A s we have seen, the legendary dodo was last seen alive on Mauritius in 1662 and is presumed to have disappeared shortly afterwards. Like the tortoises of the Galápagos, the bird had been a victim both of mankind's need for food and of the domestic animals and vermin that had flooded into Mauritius via passing ships and settlers.

The dodo became extinct in such a short space of time that only a handful of bones made their way into museum collections. As the Victorians became obsessed with the natural world, they developed an urgent need for more dodo bones but, despite much searching, Mauritius seemed to be completely devoid of them. Mauritius lacked not only live dodos but their fossilised bones. However, help was at hand.

During the 1840s the British government, which was in control of the island, embarked on a re-education programme, the aim of which was to eradicate any hint of the French culture that had existed prior to the British takeover, especially the language. As part of this, a series of teachers were imported from Great Britain. One of these happened to be George Clark, a young man with a passion for the natural world, who arrived on Mauritius in May 1851. One of his first actions was to join the Mauritius Natural History Society where he learnt not only of the fate of the dodo but of the failure to find any fossil specimens, something that the Society blamed on the island's thin volcanic soil.

By good fortune Clark had been posted to the town of Mahebourg in the south-east of the island, which lay near a swampy region known as La Mare aux Songes. Clark became fixated on the idea that this swamp was the most likely place on Mauritius to find dodo bones as it was wet and muddy – ideal for preserving delicate animal parts. Before he could investigate the area, his attention was distracted by the building of a railway so he spent much of his free time scouring the new cuttings for fossil bones.

By 1864 all thought of La Mare aux Songes was forgotten until, one day, one of Clark's pupils mentioned that some bones had been found by some manual workers cutting peat. The bones turned out to be from a giant tortoise. Clark's reaction was immediate:

> I repaired to this spot, called 'La Mare aux Sanges', and mentioned to Mr. De Bissy, proprietor of the Plaisance estate, of which this marsh forms part, my hope that, as the bones of one extinct member of the fauna of Mauritius had been found there, those of another and a much more interesting one might also turn up. He was much pleased with the suggestion, and authorised me to take anything I might find there, and to give orders to his workmen to put aside for me any bones they might find. They were then employed in digging up a sort of peat on the margin of this marsh, to be used as manure.

Nearly a year passed and no new bones appeared. Finally Clark hired some workers and sent them right into the centre of the swamp where the water was deep and the vegetation thick. He ordered his men to start feeling in the mud with their feet. A few minutes later, one waded towards him, carrying a small thin bone. Clark recognised it as being a leg bone from a dodo. At last a resting place for the dodo's remains had been found. In the ensuing months he had the swamp vegetation removed and, in time, managed to find enough dodo bones to reconstruct an entire animal. In fact, he had so many that he made quite a bit of money selling them to various museums around the world.

With all the excitement surrounding the dodo, many other extraordinary fossils from La Mare aux Songes were completely ignored, including many giant tortoise remains. Dozens of such bones had been recovered from the swamp, but as nobody really knew what to do with them, they were just piled up in a store room by the Natural History Society. Among them was a gigantic shell, pulled whole from the swamp in the late 1860s. Despite its enormous size, it lay dusty and unwanted on a shelf until 1872, when the botanist Louis Bouton took an interest in the specimen. With the blessing of the Natural History Society, he crated it up and posted it to a friend of his in England.

A Champion for the Tortoises

ON 18 OCTOBER 1872, Albert Günther, a specialist in fish and reptiles, made his usual daily journey to work in the zoological department of the British Museum, where he had recently been made assistant keeper. Aside from his pay rise, the promotion had provided him with a good deal of political clout both within the museum and in the wider academic world outside. That clout would become invaluable in the years to come as he embarked upon a relentless struggle to preserve the world's giant tortoises.

That day, when Günther reached his desk he was confronted with a large wooden crate, addressed to him personally, that had arrived earlier in the morning. It was most unexpected, as Günther had not requested or ordered any material for weeks. After clawing his way inside the crate, he unearthed from its wrappings the gigantic fossilised tortoise shell that Bouton had shipped out from Mauritius the previous month. A note accompanying the shell explained that it had come from La Mare aux Songes and recounted the circumstances of its discovery. As Günther read about the history of the tortoise shell, a sense of excitement rose within him. He knew that there was far more significance to the fossil than the Mauritius scientists had realised.

As Bouton and his colleagues were not reptile specialists, they had, on finding the giant shell, assumed it to be a Galápagos tortoise that had 'been scattered by man' to Mauritius. Günther, who did specialise in reptiles, was greatly impressed by the immensity of the shell and went through his department's collection, trying to find a comparison. He found instead that the collections were woefully inadequate, just as the young Charles Darwin had done nearly forty years previously.

Albert Günther, the man who first brought the plight of the world's
giant tortoises to the public's attention.

A visit to the library revealed that not a great deal had been written about the giant tortoise since the 1830s, in the days of the Frenchmen Bibron and Duméril. They, it will be remembered, divided the world's giant tortoises into four distinct species: *Testudo elephantina* from Aldabra; *Testudo daudinii* from the Seychelles; *Testudo peltastes* from the Mascarene Islands; and *Testudo nigrita* from the Galápagos. As Günther's new specimen from Mauritius did not seem to match any of these, he thought it was probably new to science.

This revelation forced him to dig deeper into the natural history of the animals but it rapidly became apparent that most of the first-hand descriptions of the tortoises' appearance and behaviour came not from zoologists but directly from the logs and journals of various mariners. The scientific study of the giant tortoise was obviously still in its infancy. As Günther read some of the tales of mass butchery and the endless descriptions of the way in which a tortoise could be caught, cooked and eaten, he realised that if they were going to be studied at all, it would have to be sooner rather than later.

There and then he determined to do something about the plight of the giant tortoises. He wanted to understand them scientifically and also to 'examine specimens from [all] islands, to collect their past history, to secure the future preservation of the species and to place on record the characteristics of all known forms'. This was a considerable task, given that all the known wild populations were many thousands of kilometres away. In the following weeks, Günther dashed off letters to anyone who might be able to help him with his new-found obsession. Several museums were happy to oblige with bones and shells from their collections but few could help him with first-hand knowledge of the living species. For that he had to look much further afield.

In late November another crate arrived for him from Mauritius, containing yet another tortoise but this time it was no fossil – the animal was very much alive. It had been sent by the governor of Mauritius who was writing in reply to a letter Günther had sent some weeks earlier. The governor had not only taken an interest in Günther's request for information about the giant tortoises but had actually taken the trouble to send him a live baby one through the post.

The governor confirmed that there were no native tortoises to be found on Mauritius any more but that living ones were sometimes imported from the remote island of Aldabra where they could still be found in some number (this was the origin of the baby tortoise).

'If you keep my Aldabra tortoise long enough,' wrote the governor, 'he will become gigantic, for he is the offspring of parents weighing about 500 lbs [225 kilograms] apiece.' He added that the large ones were now very rare as they were much in demand for pets and could cost as much as £10. Many islanders considered them as status symbols.

Günther greeted this new arrival with joy although it was far too small as yet to be of much scientific use. In the meantime he spent the next few months busying himself measuring, observing and describing the dead tortoise specimens that were now arriving at the British Museum from other collections in Liverpool, Glasgow and Berlin. The Paris Museum, which held Bibron and Duméril's original specimens, was the only institution to refuse Günther's request, believing that a written description would more than meet his needs. This was a setback but, as he came into possession of more and more specimens, Günther knew that he would be able to achieve his ambition of being the first person to properly examine all the world's giant tortoise species, both living and extinct.

This also meant confronting the reality of their precarious situation. Günther was only too aware that the actions of humans had forced almost all the world's giant tortoise species to the edge of extinction and, in some cases, had tipped them over it. A few years later he would write with passion on this issue:

> By the beginning of our century, the tortoises had been pretty well swept off the whole of the islands in the Indian Ocean, so that at the present time only one spot remains where they have survived in a wild state, *viz.*, the south island of the Aldabra atoll. Although only 18 miles [29 kilometres] long and about one mile [1.6 kilometres] wide, it offers by its rugged, deeply fissured surface, which is overgrown with impenetrable bush, a safe retreat to the small number of survivors.

By the end of 1873 it had become apparent to Günther that the only sizeable population of giant tortoises left in the world was on the remote Indian Ocean atoll of Aldabra, which, although uninhabited, was under the administration of the Mauritius government.

Günther was assured by the governor of Mauritius that the Aldabra tortoises were safe, as the island was too remote to be of any interest to anyone other than the odd passing ship, of which there were few. The governor also told him that no one in their right mind could possibly have any commercial interest in Aldabra, but he was forgetting that this was the Victorian era, when economic potential could be detected in almost any spare bit of ground. Within months Günther found himself in the thick of an all-encompassing fight to save the Aldabra tortoises from joining the lengthening list of extinct species.

During 1873, news of Günther's interest in giant tortoises had spread through Mauritius and the Seychelles. He began to receive letters from bored expatriates, offering anecdotes about the tortoises and advice as to how best to obtain specimens from remote Aldabra. Few of the suggestions were actually very helpful and most of them seemed to consist of rumours of individuals or boats that might be heading towards Aldabra in the near future.

Of much more interest was a letter that Günther received from Sir John Kirk in April 1874. Kirk was a well-known character in the Indian Ocean, who had participated in Dr Livingstone's Zambezi expedition and then settled down to become the British Consul for the East African coastal province of Zanzibar (now part of Tanzania). Zanzibar was a major trading port and as a consequence Kirk got to hear much of the gossip from the various British-owned colonised islands in the Indian Ocean.

The previous year, Kirk, who was himself a noted zoologist and botanist, had got wind of Günther's interest in the Aldabra tortoises but had thought little of it. Then in early 1874 he heard another rumour, which he knew would alarm Günther. On 2 April he dashed off a letter:

> My Dear Günther,
> I hear that the Mauritius government has given a cession of the Aldabra Island to a speculator who is trying to cut the timber for profit. If this speculation floats, the tortoises will soon go, so lay in a stock. They will rise in value in the next 50 years.

You know how a gang of [Africans] and a few English artisans would kill tortoises without end or object.
Yours sincerely,
  John Kirk

Kirk's letter was unusual in that, whilst apparently showing concern for the tortoises, at the same time he seemed to offer Günther a financial tip-off to stockpile tortoises before their price soared. Günther was not interested in the prospect of wealth through amassing tortoises and focused his attention instead on the more serious news that Aldabra Island, whose remoteness and inhospitable terrain had thus far saved it from development, was about to be invaded by a team of loggers. He knew only too well that all previous contact between humans and tortoises had ended in disaster for the latter. Although he was not a campaigner by nature, Günther knew that he could not sit back and watch while the last untouched population of giant tortoises vanished from the earth via a logger's cooking pot.

Within a day of his having received Kirk's letter, Günther had formulated a plan. The only way that he could see to protect the tortoises would be to mount a rescue operation, gathering as many of the creatures as possible from the wild state in Aldabra and transferring them to a special park in Mauritius. It was a bold move, which would involve establishing the world's first purpose-built preservation park (as opposed to the notoriously cramped and exhibitionist zoological parks, of which there were hundreds). It would also require a great deal of time, effort and money, none of which he had to hand. Given this, would the Mauritius government be willing to consider his plan?

His first action was to send a letter to Kirk, who was not only well informed of regional gossip but also politically well connected too. A reply came straight back: 'The place for forming a park of these strange tortoises seems to me easy and inexpensive and I think it is one the Mauritius legislature would take a pride in.'

Günther moved fast. He wanted to draft a petition to the Mauritius government, to plead the tortoises' case, but in order to do so he would need some heavyweight backing. His first port of call was his boss, the notoriously cantankerous and arrogant Richard Owen. Günther's letter to Owen ensured not only that his boss knew that he would be putting his name to a case that was already certain of success but also that the petition would be seen by several noteworthy individuals:

> I have taken every precaution to ensure success to the application. Sir Bartle Frere, Mr V. A. Williamson and Dr Kirk agree in declaring the scheme easy of being carried out. Sir Arthur Gordon, with whom I am in conversation about these tortoises, takes a great interest in them, and Mr Newton, the Colonial Secretary, of course fully understands the scientific importance of the case.

The notoriously self-centred Owen lapped up Günther's name-dropping and immediately wrote back: 'I would have much pleasure in adding my name to the memorial.'

Other big names soon followed. Sir Joseph Hooker, President of the Royal Society, signed at once, adding: 'it is perfectly monstrous that an island which it is worthwhile to hold should be left by the British Government to be plundered and ruined, turtles, trees, tortoises and all.'

Günther's old friend, the now-ageing Charles Darwin, was likewise concerned, agreeing at once to add his name. The presidents of the Royal Geographical Society and Cambridge University all followed suit. Within a month of his having been alerted to the threat to Aldabra, Günther's petition was despatched to Sir Arthur Gordon, the then governor of Mauritius. It succinctly set out the case in favour of the preservation of Aldabra:

> To His Excellency the Honourable Sir Arthur Hamilton Gordon, K.C.M.G., Governor and Commander in Chief of Mauritius and its Dependencies

> We the undersigned respectfully beg to call the attention of the Colonial Government of Mauritius to the imminent extermination of the gigantic land tortoises of the Mascarenes, commonly called 'Indian Tortoises'.

> These animals were formerly abundant in the Mauritius, Réunion, Rodrigues and other islands of the Western part of the Indian Ocean. Being highly esteemed as food, easy of capture and transport, they formed, for many years, a stable supply to ships touching at those islands for refreshments.

No means having been taken for their protection, they became extinct in nearly all the islands, and Aldabra is now the only locality where the last remains of this animal form are known to exist in a state of nature.

We have been informed that the Government of Mauritius have given a concession to a person who intends to cut the timber in Aldabra. If this project be carried out it is to be feared, nay certain, that all the tortoises remaining in this limited area will be destroyed by the workmen engaged in cutting the timber.

We would, therefore, earnestly submit it to the consideration of your excellency whether it not be practicable that the government of Mauritius should cause as many of these animals as possible to be collected before the wood cutting parties are landed with the idea of their being transferred to the Mauritius or Seychelles islands where they might be deposited in some enclosed ground, and protected as property of the Colony . . .

The rescue of these animals is recommended to the Colonial Government less on account of their utility (which now-a-days might be questioned in consideration of their diminished number, reduced size and growth, and of the greatly improved system of provisioning ships), than on account of the great scientific interest attached to them. Beside a similar tortoise on the Galápagos islands (now also fast disappearing) the tortoise of the Mascarenes is the only surviving link reminding us of those still more gigantic forms that once inhabited the Continent of India in a past geological age . . .

It is confidently hoped that the present Government and people who supported the 'Natural History Society of Mauritius' will find the means of saving a contemporary of the dodo.

London, April 1874

The petition was heartfelt, well phrased and supported by some of the largest scientific names of the day but would it have the desired effect? Before Günther could receive a reply from Arthur Gordon, another problem arose.

Victor Williamson, one of Günther's new-found correspondents in the Indian Ocean, wrote to warn him that the monsoon season was fast approaching and that, if he did not make haste, any attempt at removing the tortoises would be lost for several months while the ocean-going ships were unable to operate.

Things looked bleak. The lumber parties were scheduled to land on Aldabra at any moment and there was nothing that Günther could do about it. Even if he had had the money, he would still have to find a boat with a willing crew and then acquire some land on which to place the tortoises. The monsoon also brought severe disruption to the mailboat services: the Seychelles would cease to receive mail altogether, while the service to Mauritius would become patchy. Günther would have to sit on his hands and wait for news.

It was to be three months before he received a reply from Sir Arthur Gordon. The letter was diplomatic and began by expressing his sympathy for Günther's appeal. The governor even offered to obtain a gigantic male specimen of 300 kilograms but warned that the man who owned it had been saving the animal to be the main course at a feast. 'How much,' asked the governor, 'would you be prepared to pay for such a specimen?'

On the face of it the letter looked positive. The governor openly supported Günther but at no point he did offer any practical advice as to how the transfer of the tortoises might be accomplished, nor where they could be housed once away from Aldabra, nor how the whole venture was to be financed. In fact, the governor's offer to try to procure a large local tortoise at Günther's expense made it clear that there would be no money forthcoming from the Mauritius end of the venture.

In the time that it had taken the governor to reply, several more months had elapsed, adding to Günther's fear that while he did nothing the Aldabra tortoises were being wiped out.

Günther might have thought that his appeal had fallen on deaf ears, but in fact Governor Gordon had been rattled by the petition. The list of some of the most eminent scientists in the Western world made him uneasy. Sir Arthur was not fond of Mauritius and had, shortly after arriving there, declared that parts of it were 'hideous'. The arrival of Günther's petition coincided with his bid to move to a new position, preferably in the Pacific.

To be sure that there would be no further trouble from Günther and his co-signatories, Gordon took it upon himself to write a long explanatory letter, not to Günther but to Joseph Hooker, who, as Royal Society President, was the most powerful of the names on the petition. This letter offers several excuses as to why Gordon felt unable to do anything about the Aldabra tortoises. The first of these was rather fatalistic in tone.

[You] are not correct in supposing the island of Aldabra to have been already conceded to parties intending to cut timber there, but I believe the process of extermination is proceeding quite as rapidly as if this were the case. Not only are the animals destroyed by the whalers who from time to time land there for wood and water, but I am informed by those who have been there that the pigs which are supposed to have been left there by a passing ship some years ago, and which have rapidly multiplied, tear up the eggs in great numbers and even devour the young tortoises.

Governor Gordon then went on to explain that when the lease to Aldabra was offered, a clause would be put in to protect the tortoises, although, he added, it would probably be little regarded. If any tortoises were transferred from Aldabra to Mauritius, said Gordon, he would personally ensure that they had a paddock in the Botanic Gardens there. After explaining that some Seychellois kept the tortoises as pets in order to eat them, Governor Gordon concluded: 'it is probable that even without the intervention of the Government that this race of gigantic tortoises will for a considerable time longer be preserved in these islands.'

Again, Gordon's tone seemed positive and he certainly provided more background to the situation but he also gave the impression of a man unwilling to give much help. While he was happy to provide the land, it would be up to the signatories themselves to deal with the tortoises.

Günther despaired, believing that he was watching the fate of one of the world's most magnificent reptile species slip through his fingers. He was wrong. His petition to Gordon had had a greater effect than he had originally supposed.

～

When viewed through modern eyes, the Victorian scientific age was unusual and contradictory. Many scientists stood in awe at the magnificence of nature and yet they were also quite comfortable with the notion that wildlife could be collected randomly and indiscriminately, depleting the world of its stock of rare plants and animals.

The Victorian craze for collecting seashells, for example, left many European seashores completely devoid of all but the commonest of molluscs. The rarest species were the most highly prized (for example, the five-shilling shell was so called because of the price it fetched to collectors), rendering them rarer still. How men who valued nature could at the same time pursue their fascination to the extent of destroying it remains an anachronistic puzzle. Günther, an archetypal Victorian naturalist, was little different and, while campaigning for the cause of the Aldabra tortoises, would also help to make them more vulnerable.

In the late summer of 1874, a few months after sending his petition, Albert Günther was surprised to receive another letter from Sir Arthur Gordon. The letter came not from the Seychelles or Mauritius but from the English town of Staines, a few kilometres south of London. Gordon, who was home on leave, wanted to meet the naturalist. Furthermore, Gordon said that he had brought with him four baby Aldabra tortoises – was Günther interested in them?

Instead of telling Gordon to take the four tortoises back to Mauritius so that they could form the start of his tortoise reserve, Günther accepted them, even though they were extremely unlikely to reach adulthood, let alone breed. As with many naturalists of his generation, Günther's desire to own specimens of the animals he loved was in fact contributing to their precarious state in the wild.

In late August, Gordon made the journey across London to visit Günther at the British Museum, where the four tortoises were duly handed over. Much to their surprise the two men found that they had a great deal in common and a friendship was struck up. On a subsequent visit Günther took Gordon to the famous zoological gardens in London's Regent Park. In the 1850s the zoo had held a magnificent Aldabra tortoise, once owned by Queen Victoria, whose presence had drawn crowds from far and wide. Günther remarked on how much this giant beast, since deceased, was missed by both the zoo's staff and the visiting public.

This lament for the lost London tortoise obviously had an effect on Gordon, for a few weeks later a letter arrived for the naturalist from Seychelles House. The governor, now back at home, had been thinking about the lack of a giant tortoise in the zoological gardens and offered to obtain a 240-kilogram specimen for the princely sum of £10.

What Gordon proposed was to take a giant tortoise away from its natural surroundings and transport it to England, whose climate had claimed the lives of every tortoise ever to have arrived there. Again, instead of turning him down and asking that the specimen be used to found his cherished dream of a tortoise reserve, Günther agreed at once.

When writing the letter, Gordon already had a particular tortoise in mind but it was not going to be easy to get access to it. The animal had been imported into the Seychelles from Aldabra some sixty years previously and had changed hands a few times since then. The tortoise wound up in the care of a local man who kept it penned in his garden, feeding it banana leaves and pumpkins, but without the slightest interest in it as anything other than a status symbol. (To the Seychellois, having a giant tortoise in the garden was the equivalent of having a Ferrari parked on the drive.) When initially approached by Gordon, the owner announced that it was his intention to keep the enormous beast alive until one of his daughters wished to marry, when it would be served as the central course of the wedding feast. However, in a poor island money talks and it was not long before the governor persuaded the local man to part with his gigantic pet.

In fact, when Gordon wrote to Günther in May 1875 he revealed that he had managed somehow to obtain two tortoises, a 350-kilogram male (the one initially intended for the wedding feast), and also a 240-kilogram female whose origins were unknown. Both were now residing in the gardens of Government House, awaiting instructions.

To seal the deal, Gordon included with his letter a photograph of each tortoise. One of the photographs showed a local boy sitting on the male tortoise, dwarfed by the size of its shell. This was all the encouragement that Günther needed. He replied at once, asking that the tortoises be despatched on the next available ship.

The governor's secretary wrote back to confirm that the arrangements were all in place. The tortoises would depart for England, via

Marseilles, on 6 June. However, as Günther read on, it became clear to him that moving the tortoises presented more logistical difficulties than he could possibly have envisaged.

The animals, wrote the secretary, had been placed in two purpose-built wooden cages but as the large male had already made one escape attempt, the cage had had to be reinforced. 'His strength is prodigious,' commented the secretary.

The tortoises would be accompanied by a watertight tank and four barrels of rum so that, should either of them die, the body could be preserved. Their food for the journey was two sacks of dried leaves, two sacks of banana leaves, two sacks of sweet potato tops and no less than nineteen pumpkins. The large tortoise in its cage weighed over one tonne on its own, and the weight of all the other paraphernalia came to about the same again. At every stage the secretary's letter stressed the amount of time, materials and manpower that had been sunk into the venture. His reason for doing so became obvious at the end where there appeared a neat account of the costs to date. Günther read this with horror.

Shipment of Two Tortoises	£40	0s	0d
Material and Construction of Two Cages	£7	7s	3d
Material and Construction of Watertight Can	£3	9s	5d
Rope Sling for Cages		17s	5d
Rum for 4 casks	£39	2s	8d
4 Casks	£4	4s	0d
Total	**£94**	**0s**	**10d**

When he added on the £10 that he had initially paid for the tortoises, Günther suddenly found himself saddled with a bill of over £100, a small fortune in those days.

Although he had had ambitions of letting the tortoises reside in the zoological gardens, he had not actually approached the Zoological Society on the matter of payment as he had not expected the costs to climb to such a level. For a poorly paid keeper in the British Museum, finding £100 was no easy task. Furthermore, Gordon's secretary had asked for prompt payment.

Günther was forced to go cap in hand to the Zoological Society, which was not best pleased to receive him. When he first mentioned the idea of the Society housing and feeding the tortoises, this met with no opposition, but he had neglected to mention that the Society would have to pay for the animals' transport as well. In truth, Günther had thought that he himself would be able to cover the cost, not suspecting what a can of worms he was opening.

At first the Zoological Society was very reluctant to have anything to do with Günther's added expense but the gentle scientist reminded the Society of the popularity of its last tortoise. Did not Queen Victoria herself express her admiration for the beast? Finally, Philip Sclater, the Secretary of the Zoological Society, agreed to ask the Purchase of Animals Committee to buy the tortoises on 'the understanding that the Trustees of the British Museum take them off our hands at the same price when they die'.

Günther reluctantly agreed to the terms but he knew that the British Museum's Trustees would be outraged by his having made such a large financial deal behind their backs so he reluctantly added: 'in case of difficulty arising, I should hold myself personally responsible to repay to the Society the purchase money.'

Günther duly approached the museum's Trustees, told them what he had done and stood back to wait for the blast. It came a few weeks later in a letter that expressed their 'severe displeasure' at the deal made on the museum's behalf. One hundred pounds was a great deal of money to pay for two dead tortoises and, to make matters worse, the corpses were to be obtained second-hand from their academic rivals at the Zoological Society! Günther was not in the museum's good books.

Matters improved on 4 July when news reached him that the tortoises had arrived at London's Victoria Docks and that they were alive and well. Günther celebrated but was dismayed to learn that the four casks of rum, which had travelled with the tortoises and which cost almost £40, had mysteriously gone missing. This was a blow as he had intended to sell the spirits to offset the cost of the shipment.

After he made a few enquiries, the casks turned up in Marseilles, which, owing to an error on the shipment form, was listed as their final destination. Even though the error was obvious, French

Customs had impounded the rum, pending the payment of an import duty of threepence a gallon. No amount of reasoning would sway the port officials and so reluctantly Günther paid the tax, complaining all the while that he was out of pocket because of somebody else's mistake.

Worse was to come. A deal had been struck with a distillery, Widenmann, Broichu & Co., which had agreed to take the rum at one shilling and sixpence a gallon for methylating. However, when the casks finally entered the country the distillers wrote to Günther to inform him that the rum was weaker than the strength stated on the label. Someone, they said, had diluted the spirits with water. It looked as though some of the sailors on the long voyage (or possibly the French Customs officials) had been helping themselves to the cargo, replacing what they had stolen with water. The price paid by the distillers was reduced by threepence a gallon.

The large reinforced cages did at least fetch £8 so that the total cost of the venture was now reduced from £110 5s 9d to £92 0s 0d. The reduced figure still did not amuse the museum Trustees but they were used to their scientists making illicit financial deals and, in the scheme of things, Günther's was relatively minor, compared to some of the covert transactions over the years. After all, there was always the chance that the tortoises would live to their full term of 180 years or more, by which time inflation would have reduced the debt to practically nothing.

Before being passed to the Zoo, the tortoises rested briefly at Günther's home in the South London suburb of Surbiton. Later Günther's son would write that one of his 'most cherished of boyhood memories was being lifted up for a ride on the backs of two giant tortoises that [my father] kept for a short time in his tiny garden'.

Finally the time came for the two giants to be delivered to their new quarters at the London Zoo. The *Illustrated London News* and *The Times*, among others, covered the event in full detail, ensuring that within a short time it was apparent that all the effort had been worth it. The tortoises became an instant tourist attraction, gaining the Zoo both publicity and increased visitor numbers. Londoners talked of going to see 'the big tortoise', which was the male, and his smaller

This Aldabra tortoise was bought, at great expense, by Albert Günther and shipped to England. It died soon afterwards.

female companion, which for some reason was often referred to as being 'sulky'. Günther received much admiration from his friends as well. Joseph Hooker wrote, congratulating him on the venture and offering the full disposal of Kew Gardens' resources.

The two tortoises, but especially the big male, were famous across the world. One New York correspondent wrote of his visit to London Zoo:

> You may have a look at the lordly lions, and the restless tigers, or the lower feline creatures who are merely fidgety; you may only just glance at the monkeys, and be merely indistinctly aware of those distressful prisoners, the bears and by doing so you may have, at least, a good general notion of what these creatures are like. But the big tortoise cannot be taken in at a glance; no one with the most ordinary perception of the fitness of things would propose to 'have a look' at him, or if he did, he would feel the impropriety of the expression, when the quaintness, the hugeness, the incongruity, and the ancientness, retrospective and prospective, of the creature should have revealed themselves with his first glimpse of him. There is, however, no such thing; he is too big and too slow for a first glimpse.

Based on eulogies such as this it is quite probable that the London Zoo reclaimed the money spent on the tortoises within a relatively short period of time. This was perhaps just as well, because the English climate did not agree with the larger tortoise and, after its first winter, it began to show signs of distress, losing weight and becoming inactive and withdrawn. By the autumn of 1876 the world's largest known tortoise had died. Had it been left alone in its owner's tropic garden, it would doubtless have lived although, admittedly, there was still the danger of its becoming the central course at a wedding feast. What became of the female tortoise at the zoo is not known.

The buying and sending of the two tortoises was one of the last actions taken by Sir Arthur Gordon as governor of the Seychelles and Mauritius islands. He had for some time been pressing for a commission to the Pacific and in the spring of 1875 his wish came true when he was made the first official governor of Fiji. He left the Seychelles only a matter of days before the two tortoises, sailing off in the opposite direction.

Gordon's brief flirtation with the giant tortoises over which he had political charge had ended but his legacy would live on. Günther's enthusiasm and quiet persistence had paid off. One of the departing governor's last instructions was to ensure that the Aldabra tortoises be safeguarded from the ravages of man. These were followed to the letter and, on 12 November 1875, the new governor declared that the giant land tortoises on Aldabra should be a protected species.

The good news reached Günther at an opportune time for he had just finished one of the most comprehensive studies ever made of the giant tortoise and was now ready to announce his findings to the world.

Understanding the Tortoises

WHEN GÜNTHER FIRST settled down to study the world's giant tortoises, some sixty-odd years had passed since August Schweigger had examined a single giant tortoise shell and given it the name *Testudo gigantea*. In the meantime a few other scientists had taken their own turn at naming new giant tortoise species, muddying the waters considerably.

Unfortunately, most of the new names had been given by people who had only one or two specimens in their possession and thus could not compare them with any of the other tortoise specimens worldwide. By the time that Günther came to look into the problem, he was dealing with a jumbled mass of more than a dozen different scientific names that had been applied to only a handful of tortoises. It was inevitable that mistakes had been made and that some tortoise species had been afforded two or more names. To that date no one had been overly worried about these mistakes, as the only way to rectify them was to spend a great deal of time and effort collecting original specimens, followed by hours of comparing them in the smallest of detail.

Hitherto nobody had bothered to undertake this task but Günther realised that, in order to be able to assess the true status of the world's living giant tortoises, he first had to know just how many different species there were. It was with this aim in mind that he began gathering together giant tortoise specimens from across the world.

By 1875, Albert Günther had not only obtained a guarantee that the Aldabra tortoise was a protected species, but he had also managed to get his hands on over forty giant tortoise specimens, both alive and dead, from various museums in Europe as well as from generous individuals such as Sir Arthur Gordon. Once surrounded by this profusion of animals, he set about trying to distinguish one from another, separating them into discernible species. This task was far more onerous than might at first appear.

For a start, many of the specimens were not accompanied by precise information about where and when they were captured, making it difficult not only to tell which island they belonged to but also exactly where on that island. Like Darwin before him, Günther knew that because most tortoise species were restricted to particular islands, an exact island locality was necessary in the speciation process. Then there was the problem of immature specimens whose shells and bones had yet to develop their true adult physical traits. A number of young tortoises had been described as new species but in all likelihood they were probably just juvenile versions of adult tortoises that had already been given names. How was Günther to resolve this?

The answer is by painstaking observation, measurement and comparison. Projects of this kind can really only be undertaken by those with a certain temperament. They must be cool, calm and be able to commit themselves to hours of tedious, repetitive and mind-numbingly boring work. Measurements must be taken down to the last millimetre. Each bone or shell fragment must be examined and described to its last lump, bump or scratch. Although such projects are extremely beneficial to science, few undertake them voluntarily. By good fortune Günther was blessed with the necessary temperament for the task and worked on his specimens for over a year before feeling able to pronounce on the world's giant tortoises.

The announcement came in a fifty-eight-page article that was published in the *Proceedings of the Royal Society* in 1875. It was the first time that all the giant tortoise species had been examined collectively and organised into some form of coherent order. Günther's results were to be the start of a new era in the history of the study of the giant tortoises. He began by scotching an old myth:

Under the name *Testudo indica* were comprised all gigantic land-tortoises brought to Europe in ships which, on their return from India, had touched at the Mascarenes. When, at a later period, zoologists became acquainted with a similar tortoise from the Galápagos Islands, some considered the latter specifically distinct, whilst others maintained that they were specimens of the same species, which had been scattered by man in different tropical parts of the globe. However, a closer examination and comparison of the remains at my disposal revealed important differences unmistakably pointing to a multiplicity of species. The results of these researches were startling, and may arrest the attention of the zoologist, all the more as the facts elucidated bring us face to face with the mystery of the birth and life of an animal type.

Günther then laid out his conclusions. His first was to dispel the notion that there was only one giant tortoise species in the whole world. There are, he stated, two general populations of giant tortoise: one restricted to the Indian Ocean islands, the other to the Galápagos Islands. The morphological distinctions between the two populations are clear. Never again should the world's giant tortoise population be referred to under the single name of *Testudo indica*.

His second was to state that, within the tortoise populations of the Indian Ocean and Galápagos Islands, it was possible to detect a number of distinctive species. In the Indian Ocean there were discernible species on Mauritius, Rodriguez and Aldabra, while Günther believed that in the Galápagos Islands he could identify at least five species, each of which was restricted to a particular island.

This last point, which confirmed what Darwin had suspected over thirty years previously, was apparently a vindication of his theory of natural selection. However, although Charles Darwin and Albert were firm friends, Günther was not a supporter of natural selection, preferring to believe that evolution was guided by divine providence rather than random mutations. Throughout his many published works on giant tortoises, Günther did not once allude to the principles of natural selection even though many of his findings provide clear evidence in favour of it. Unlike Darwin, he did not have a wider point to prove with his study of the tortoises. He was simply stating the facts as he saw them. It would be up to others to interpret his results in terms of a wider scientific philosophy.

Günther had divided the once all-encompassing nomenclature of *Testudo indica* into at least eight species, three in the Indian Ocean and five in the Galápagos, but there was still much work to be done. He admitted that poor locality information meant that he could not be certain which island each of the new Galápagos species came from. He also reported that of the Indian Ocean species, only one, the Aldabra tortoise, was still living in a wild state.

Günther's 1875 monograph was to be the foundation for a new era of research on the giant tortoises. He had stamped out some of the myths that had been rattling around the scientific community for over half a century. Because of his careful observations and measurements, there could now be no doubt that the giant tortoises were more diverse and endemic than anyone, other than perhaps Darwin, had hitherto expected.

Despite this progress, Günther's work on the tortoises was only just beginning. Although he had distinguished discrete tortoise populations and identified some of their species, he knew that there were likely to be other species as yet undiscovered. He did not pause to receive the deserved praise that poured in for his monograph, returning instead to his specimens and pursuing the acquisition of new ones in order to better understand the animals with which he had become obsessed.

The world would have to wait another two years before Günther published again. This time it was not a small monograph but a hundred-page book entitled *The Gigantic Land-Tortoises (Living and Extinct) in the Collection of the British Museum.* From the outset it was obvious that this would remain the standard work on giant tortoises for many decades.

Günther's obsessive study had revised his ideas of only a couple of years previously. Instead of there being two distinctive giant tortoise populations, he now believed that there were three. Those of the Galápagos Islands, those of the Mascarene Islands (Mauritius, Réunion and Rodriguez) and those of Aldabra and the Seychelles. Based on his observations, he judged that it was possible to discern not eight species of tortoise, as he had originally thought, but twelve, and for the first time he felt confident enough to attach locations to these species.

Günther's list of tortoise species was as follows:

THE RACES OF THE ALDABRA GROUP
Testudo elephantine
Testudo daudinii
Testudo ponderosa
Testudo hololissa

THE EXTINCT RACES OF THE MASCARENES
Testudo triserrata (Mauritius)
Testudo inepta (Mauritius)
Testudo leptocnemis (Réunion)
Testudo vosmaeri (Rodriguez)

THE RACES OF THE GALÁPAGOS
Testudo elephantopus (San Salvador Island?)
Testudo nigrita (?)
Testudo vicina (the south of Isabela Island)
Testudo microphyes (the north of Isabela Island)
Testudo ephippium (Santa María Island)
Testudo abingdoni (Pinta Island)

Again, despite the obvious evolutionary implication of so many tortoise species being restricted to individual islands, Günther did not use his new classification to support Darwin's big idea. In fact, he tried his hardest to deny it.

Darwin's firm belief was that animals and plants evolve through spontaneous mutations that accumulate over generations to change slowly the appearance of an organism until it differs enough to be called a new species. Much of Darwin's *prima facie* evidence had been taken from remote islands such as the Galápagos where the isolation of plants and animals appeared to have caused them to evolve into new forms that were perfectly adapted to the environment in which they were living.

Taking the case of the Galápagos tortoises, it was Darwin's belief that thousands of years ago a species of tortoise had been swept away from the South American coast and had by chance been carried by the currents to the Galápagos where, miraculously, it was beached alive. The tortoises (perhaps originally a pregnant female or even a male and female washed away together) then became established on the Galápagos and, in the absence of any predators, grew large in order to be able to store enough fat and water to see them through the nightmarish drought conditions that frequently prevailed in the islands. Some of these large tortoises were themselves swept away to other islands in the Galápagos where they too evolved to adapt to local conditions. By the time the first humans landed on the Galápagos, the giant tortoise populations had been isolated for so long that each breed of island tortoise had evolved into a new species (except on Isabela where there were, according to Günther, at least two isolated tortoise populations, in the north and south). The same process was thought to have happened in the Indian Ocean except that the original tortoise was believed to have originated on the African or Indian mainland.

This position was the standard Darwinian line on the origin of the giant tortoises. The fact that there were populations of giant tortoises in both the Galápagos and the Indian Ocean islands was simply coincidence and a result of the lack of predators on the respective islands. The bottom line was that all the world's giant tortoises were Robinson Crusoe-style castaways and that, fundamentally, there could not be any direct connection between the two main populations.

Because of the chaos over the number of species and their relationship to the islands on which they lived, the giant tortoises had not been much of a celebrated cause among Darwinists. Many, including Darwin himself, suspected that there were numerous species and that each would be restricted to an individual island, but until Günther's 1877 paper they did not have the necessary proof of this.

Günther himself was well aware that his work could be used to bolster the theory of natural selection to which he was virulently opposed, so he took the decision to head this off by suggesting a theory of his own. In doing so he was opening up a great debate about the origins of the giant tortoises that would rattle on for well over a century.

If Günther was to dispel Darwin's theory, he would need to be able to prove two things: (1) that the giant tortoises did not evolve on the islands on which they are (or were) currently found to be living; (2) that there is a direct genetic link between the giant tortoises of the Galápagos Islands and those on the various islands in the Indian Ocean. This was not an easy task but Günther felt sure that he had a solution to the problem and he chose a public meeting in London on 26 January 1877 to air his views.

In front of many learned colleagues, he first outlined his belief that whilst there were obvious differences between the Galápagos and Indian Ocean tortoise populations, there were also certain similarities that suggested that the two groups of animals were closely related to one another.

'How,' asked Günther, 'could this be explained with the aid of the doctrine of either a common or manifold origin of animal forms?' In his mind the two populations could not have evolved independently of one another and must once have been linked. He continued:

> The giant tortoises must once have been spread across the entire globe but the arrival of man spelt the start of their extinction. Hunters would eagerly seek them, as they were the easiest of his captures, yielding a most plentiful supply of food. Consequently the tortoises were exterminated on the continents, only some remnants being saved by having retired to places which, through the submergence of land bridges, had become separated from the mainland before their enemies could follow them.

This radical theory implied that the world was once awash with giant tortoises that had been mercilessly hunted to extinction by Palaeolithic man. Some of these tortoises, according to Günther, had been lucky in having crossed thin land bridges to remote islands, which, after these bridges had disappeared, kept them out of harm's way until the arrival of the European sailing ships.

Günther's dismissal of Darwinism did not convince everyone. In the audience that night was Sir Joseph Hooker, the best friend of Charles Darwin. He could see an instant problem in Günther's way of thinking and the next day wrote to his mentor: 'Last night we were party to Günther's fine paper on the big tortoise. He speculated widely about a land connection between the Africa Islands and continental Africa and this with the continent of America and the Galápagos to account for the affinity of the Mascarene and Galápagos beasts. Oddly enough it never occurred to him that the said tortoises all inhabit volcanic islands; which when I pointed it out, he denounced.'

Thomas Huxley, also known as 'Darwin's bulldog', drew attention to the impracticability of there having been land bridges several hundred kilometres in length joining isolated islands but it was all to no avail. Günther's mind was made up. In years to come the scientific community would be split over the origins of the Galápagos and other island tortoises.

When Darwin heard of Günther's theory, he was, as ever, philosophical about it. In a letter to Hooker he commented, 'How the deuce they got to volcanic islands I cannot pretend to say.'

Having said his piece, Günther relinquished the issue of the origin of the tortoises and focused on other matters. He never did convert to Darwinism and it must be assumed that he held on to his land-bridge theory until his death.

The issue of the origin of the tortoises was to erupt again a few years later, after the eminent American scientist Alexander Agassiz visited the Galápagos Islands and came to the firm conclusion that they were volcanic in origin, which suggested that the tortoises must have travelled there by sea. This brought forth a violent riposte from George Baur, a fellow scientist and tortoise expert, who supported the idea of there having been a land bridge between the Galápagos and the South American mainland. The two men slugged it out with one another in the scientific press, accusing each other of deliberately misquoting their papers or of purposefully misrepresenting their opponent's arguments.

As time progressed and accurate measurements were made of the sea floor around the Galápagos, it became obvious that the water surrounding and between the islands was exceedingly deep (several kilometres in places) and that no submerged land bridges could possibly have existed on the seabed around the islands. With this option gone, it was suggested by some people that the animals of the Galápagos had been bodily picked up from South America by tornadoes and then dropped on their respective islands. By the 1920s all but the most stubborn individuals accepted that the Galápagos, Mascarene and Aldabra islands were all volcanic in origin. This meant that the tortoises and other island wildlife must have been carried to the islands only after each archipelago had risen from the sea, adding a great deal of weight to the Darwinian argument.

Günther was not particularly troubled. His 1877 book was to be his last written contribution on the subject of tortoises for some considerable time, and he was not prepared to get involved in a debate over the origin of the islands and their fauna. He was more concerned with trying to drag back from the edge of extinction those giant tortoises that were still left alive in the world.

In 1875 he believed that he had scored a major victory in having gained assurances on the future of the Aldabra tortoise and that he could take a well-earned rest, but he was wrong. In the years ahead the tortoises were going to need all the help they could get.

Save the Tortoises!

A FTER THE PUBLICATION of his 1877 magnum opus on the plight and scientific status of the world's giant tortoises, Albert Günther was able to take things easy for a while. Indeed, ill health (apparently brought on by stress through overwork) forced him to ease the pace. Occasionally a parcel would arrive at the museum, containing the bones or other remains of a recently deceased (or sometimes long-deceased) tortoise. Thanks to his campaigning and monographs, Günther's notoriety as *the* expert in giant tortoises had spread worldwide and, in truth, he was pleased to receive the parcels. The issue of the perilous status of the Aldabra tortoise was, however, never far from his mind.

In 1878, less than three years after giving assurances that Aldabra was a protected area, the Seychelles government commissioned one Sergeant Rivers to go to the island to assess its commercial viability. He reported that there was plenty of scope for a year-round settlement and that 'there is plenty of good ground on which food stuffs, including goats, might be grown.' Weary of the attention that Aldabra had already received from the scientific community, Rivers added: 'If this [settlement] were done, a sufficient revenue might accrue to pay for a guardian to protect the tortoises.'

Many a politician has made promises when trying to influence an unpopular or awkward decision but, once they have achieved their aim, few of those promises ever actually come to fruition. Nowhere was this more true than in this case. When Günther got wind of Sergeant Rivers' report, he recognised that Governor Gordon's promise was being broken and demanded an explanation.

The Chief Commissioner of the Seychelles wrote back, explaining, as Sir Arthur Gordon had first done years previously, that the giant tortoises were already on the fast track to extinction. 'Many vessels call at Aldabra for the sake of capturing land tortoises,' explained the commissioner, 'and the only way to save them from eventual annihilation is to have a government guardian on the island and to strictly preserve the forest.' Such a guardian could only be funded via the economic development of the island.

It was a classic catch-22 situation. Leave the tortoises alone and they would be eaten by passing sailors, just as had happened in the Galápagos. Yet to protect them would involve developing the island and that would mean destroying their fragile habitat (especially if herds of hungry omnivorous goats were imported), which would also lead to their eventual destruction, again, just as had happened on the Galápagos. There was little that Günther could do other than protest against the settlement plan, which, in his opinion, gave the tortoises the least chance of survival.

Fortunately the Seychelles government was having trouble finding anyone willing to develop Aldabra but the odds were that it was only a matter of time. All Günther could do was to hope that he would hear of it before anyone else so that he could halt any development plans in their tracks.

By now his campaigning had become a serious thorn in the side of the Seychelles and Mauritius authorities, whose every plan for Aldabra somehow travelled back to Günther's ears via a network of fellow tortoise obsessives in the area of the Indian Ocean.

Perhaps the most extraordinary of Günther's spies was the eminent Colonel Charles Gordon (no relation to the previous governor of Mauritius), a member of the minor landed gentry whose military exploits had been lauded around the world.

In 1881, Colonel Gordon wrote to Günther from Jerusalem to let him know that he had undertaken a survey of the Seychelles. He noted that no less than thirteen families kept tortoises as pets and then went into lurid detail about the sex lives of these reptiles. He ended by observing that 'the tortoises never quarrel among themselves.'

This letter is bizarre. It is by no means certain that Gordon ever reached the Seychelles, let alone spent hours watching tortoises mating and laying eggs (he credits most of the information to a Seychellois by the name of Charles Button). Quite where Gordon's interest in giant tortoises came from is unknown but it appeared that this famous military campaigner had somehow discovered the same passion that had gripped Günther nearly a decade earlier.

A follow-up letter from Gordon came in 1883, this time addressed to William Dyer of Kew Gardens and sent from Jerusalem, where Gordon was taking a year's sabbatical. Hooker quickly passed the letter on to Günther, knowing that it contained the news that he had been dreading.

> 5th February, 1883
>
> My Dear Mr Dyer,
>
> Mr Button of the Seychelles tells me that the Mauritius Government have let the Aldabra isle to a Mr Couvin for seven years. If you could get some warning to the Mauritius Government to see that the headquarters of the gigantic tortoises is not ruined, it would be a good thing.
>
> Believe me, with kind regards to Sir J. Hooker.
>
> Yours sincerely,
>
> C.E. Gordon

As soon as Günther heard the news he was once more activated into letter-writing mode, this time appealing to the new governor of Mauritius, Sir John Pope-Hennessy. Yet again he found himself anxiously drawing the attention of another governor of Mauritius to the plight of Aldabra. He enclosed a revised copy of the 1874 petition, this time signed by the Presidents of the Royal Society, the Linnaean Society and the Zoological Society, as well as by Lord Walsingham and Sir Joseph Hooker. A copy of his 1877 monograph was also included. In his accompanying letter Günther pleads with the governor: 'Having been informed that Aldabra is to be let, or recently has been let, to a private party, I beg leave to express the hope that means will be taken for the continuance of the protection that was accorded to the few remaining individuals of this highly interesting animal by two of the former governors of Mauritius.'

As it had done to some of his predecessors, the list of eminent names prompted the governor to react at once. Günther immediately received a reply in which the governor stated that he would not 'fail to bear in mind your works as to endeavouring to preserve the [tortoises] of Aldabra'.

The scientist had heard these assurances before, only to find that his beloved Aldabra tortoises were again under threat when the next governor took charge. This time, however, Governor Pope-Hennessy did take matters a little further. With no wish to be seen as the man that rendered extinct the last sizeable population of Indian Ocean giant tortoises, Pope-Hennessy ordered his secretary to learn the whereabouts of all the Aldabra tortoises on Mauritius and the Seychelles. Most were isolated animals kept as pets or mascots by various institutions, individuals or families. The exception, however, was the Mauritius Acclimatisation Society (an organisation dedicated to introducing alien plant and animal species to Mauritius – another Victorian bad habit), which had gathered a number of Aldabra tortoises over the years.

The political weight of the governor's office was brought to bear upon them to release their tortoises into the care of the authorities. The plan was probably to place the animals in the Botanical Gardens, as originally agreed with Günther by Sir Arthur Gordon. But the Society was stubborn. The only condition under which they would release their tortoises was if the chosen site was Flat Island, a small uninhabited rock lying off the north coast of Mauritius. Here the tortoises could live in harmony, without interference from either the governor or passing ships. Reluctantly the governor's office agreed.

A letter was sent to the governor, which explained in detail the actions that had been taken to track down the location of Aldabra tortoises in Mauritius and the setting up of the Flat Rock reserve. Aware that the eyes of the world's scientific community were upon him, the governor decided to maximise his actions by publishing the letter, which had actually been penned by one W. Littleton, his secretary, to *Nature*, one of the world's foremost scientific journals.

The letter painfully details the number of Aldabra tortoises in Mauritius, and comments that those released on Flat Rock are 'completely at liberty, that they feed themselves and are apparently doing well'. (It should be noted that I have not been able to find out what happened to the Flat Island tortoises. In 1893 it was reported that they were 'thriving, if not multiplying'. However, the island is currently devoid of these animals, indicating that they either died out or were at some point removed.) The letter ends with a typical underling's apology to his boss:

> I am sorry not to have been able to collect for your Excellency's information more details of these creatures; but I have stated enough to show that there are many specimens well known and in good keeping.
>
> I have also been unable to ascertain whether there are any of large size known to remain on Aldabra Island; but I am told that it is supposed there are in the thick scrub of the interior.

It is clear that Governor Pope-Hennessy (or at least his secretary) had gone to a great deal of effort to be seen doing the right thing. Of course, the wild tortoises in Aldabra were no better off than they had been before but that had never been the object of the exercise. The governor merely wanted to prove that, even if the wild individuals on Aldabra all disappeared, then at least the species itself would not become extinct.

Günther was not happy but recognised that this was the best result he was likely to get from Mauritius in the short term. Being stuck in London with the knowledge that an island full of wild tortoises was being ravished must have given rise to the same feelings of impotence that many modern environmental campaigners experience when, powerless to act, they are forced to stand back and watch bulldozers move in on a patch of virgin woodland. The matter became dormant again, but not for long.

The Final Showdown

THE INEVITABLE NEWS that Aldabra had been leased for commercial activities came to Günther in a note from the Seychelles authorities in August 1891. It informed him that Aldabra Island was now in the hands of a Mr James Spurs but that, in their opinion, he was 'unlikely to kill the goose that lays the golden eggs by exhibiting the rapaciousness which has characterised the actions of others who have been there before him'.

Whether the 'others' mentioned were the island's previous leasers or simply visiting sailors is not made clear. Either way, it was obvious that the authorities were perfectly aware that Aldabra had suffered considerable ecological abuse over the years, almost certainly to the detriment of the giant tortoise population. It was to be the start of a turbulent time for the Aldabra tortoises.

After decades of deliberately ignoring the remote atoll, for some reason the Seychelles government suddenly decided to take an interest in Aldabra. In May 1892, Mr T. Riseley Griffith, an emissary, was sent there on the navy ship HMS *Redbreast*. He was much heartened by what he found. Not only was the island in good health but, he wrote: 'When Mr Spurs first went to Aldabra he was of the opinion that there were very few [tortoises] left, but he now states that there cannot be less than 1000 in all the island. I made him repeat this statement more than once as I was sceptical about so large a number but he assured me that a few hundred would not accurately describe their number.'

By now much of Günther's once-extensive network of contacts in the Indian Ocean region had withered and consequently it took several months for news of the 1892 Seychelles colonial report to reach him, but he was not much cheered by Mr Spurs' optimism over the tortoise numbers. He wrote later: 'The information that I had gathered was so much at odds with that of Mr Spurs,' but, as usual, there was little he could do about it.

In the meantime, he had become embroiled in a public argument about the origins of a pet tortoise in Mauritius, which required him to write many letters. On 12 January 1893, in one of his letters to *The Times* newspaper, Günther publicly aired his beliefs about the fate of the Aldabra tortoise:

> '[The Aldabra Islands] are now leased by the Mauritian Government to persons cutting wood. And, although the protection to the tortoises is included in the conditions of the lease, there is no guarantee that the condition is kept. It is not possible to prevent labourers working in the bush from killing the young or collecting the eggs and adding them to their diet of salt fish. The only plan of saving these animals from utter extermination – and I hope that the Colonial Office and the Government of Mauritius will see their way to adopt it – is to transport every individual, old or young, from Aldabra to Mauritius, where they could be placed in a suitable compound and maintained under strict supervision.
> I am your obedient servant,
> A. Günther

It was a supreme irony that, after years of direct campaigning, this indirect passing comment would finally bring about what he had sought for nearly two decades.

A few days later Günther received a letter postmarked Downing Street, the working seat of the incumbent British government. It turned out to be from one Edward Wingfield, Secretary to Her Majesty's Government. Mr Wingfield commented that the Marquis de Ripon had seen Günther's letter in *The Times* and was moved by it.

As a Liberal MP and Colonial Secretary, the marquis was proposing to write to the Seychelles administration, suggesting that they might follow Günther's advice to take some tortoise specimens from Aldabra to the Seychelles for safe keeping.

To say that the Marquis de Ripon was a powerful man is an understatement. This senior politician had, in his time, been First Lord of the Admiralty, Viceroy of India and a grand master of the Freemasons. As Colonial Secretary, the marquis had control over all the British colonies, including the islands of the Indian Ocean. If the marquis asked the Mauritius governor to jump, his reply would be, 'How high?'

Having the political weight of the marquis behind the tortoise campaign was a once-in-a-lifetime opportunity. Günther quickly fired off a reply, in the hope that he could use the marquis' interest to drive through some of his ideas on the preservation of the tortoises. He was, he told the marquis, concerned that the large number of tortoises mentioned by Mr Griffith in the 1892 colonial report might have been exaggerated. If the marquis was writing to the Seychelles administration, could he also please send a line to the governor of Mauritius to ask him (a) whether he could confirm that there were indeed 1,000 or more tortoises on Aldabra and (b) what active steps they had taken to preserve the animals?

Having been fobbed off for years by a long list of Mauritius governors, Günther now had one real chance to put pressure on the authorities in charge of these faraway tropical paradises. With the marquis on his side, he had the Mauritius governor between a rock and a hard place.

Neither the marquis nor Günther had to wait long for a reply. It came directly from the governor, now one Hubert Jerningham, on 4 April 1893, and it confirmed much of what Günther knew already: 'Mr Spurs' estimate of [the tortoises'] numbers was considerably over-stated, but Mr Spurs has at all times issued the strictest instructions to his people at Aldabra never to take or kill any of the land tortoises there and I have no reason to suppose that they are destroyed by his labourers. Their greatest enemy is [not man but] rats who doubtless destroy eggs.'

The pattern was a familiar one. Yes, the tortoises were in trouble but there was nothing the governor could do and, anyway, whoever leased the island was not to blame. So far, three different governors had laid responsibility for the demise of the tortoises at the door of pigs, passing ships and now rats. Each of the governors had also, as an act of benign generosity, offered to produce some local tortoises and house them somewhere; Hubert Jerningham was to be no different. If, he wrote, the UK government could provide him with 400 rupees, he could buy two giant tortoises (currently in private hands) and ship them to the Seychelles island of Curieuse, where boats never landed and fishing within the reef was banned.

Günther wrote back to say that he liked the Curieuse plan but commented that more than two tortoises would be needed for a viable breeding population. When he heard this, the Mauritius governor complained that tortoises could not be moved in quantity from Aldabra because they were too heavy and awkward to carry to the ships.

Again, the Marquis de Ripon seems to have intervened, relaying with some force an instruction to find more tortoises on islands in the Indian Ocean, for within months another twenty-three had been bought at auction (for 103 rupees – a bargain compared with the 400 rupees paid for the two large ones) by the Mauritius governor. Within a year, the total had reached forty-two, all of which were ceremonially taken to Curieuse Island and released into the wild.

Creating a wildlife reserve for the tortoises was the very solution that Günther had suggested twenty-five years earlier. It was not ideal but he was a realist and, although he would have liked Aldabra to be declared off limits, he knew that this was unlikely ever to occur. It had taken decades of unrelenting canvassing and campaigning to achieve his original ambition and it was, within his lifetime, as good a deal for the tortoises as he was ever going to get. A profuse letter of thanks was written to the marquis. The fate of the wild Aldabra tortoise might still be hanging in the balance but there would at least be other islands where wild populations could continue to roam free.

The Curieuse experiment was not quite the end of the story. Although his advancing age and the pressure of work forced Günther to scale down his efforts on behalf of the giant tortoises, he continued to campaign for the protection of Aldabra. Again, all his work seemed to have paid off when, in the spring of 1899, Joseph Chamberlain (the British Secretary for the Colonies) wrote to him, confirming that he agreed 'that the killing of the tortoises and their exploitation might properly be prohibited and restricted by law and [I] propose to instruct the administration of the Seychelles to introduce an ordinance for that purpose'.

The law was duly passed but, like the assurances of several other Mauritius governors, there was not the means, the money or the will to enforce it. In fact, the new law became merely academic when, in 1903, the Seychelles became a Crown Colony. This meant that the islands now had autonomy and no longer came under the political administration of Mauritius. At the same time Aldabra, which had for some years been administered by civil servants on both Mauritius and the Seychelles, was officially judged to be within the Seychelles' territorial waters. The loss of Aldabra did not lead to the shedding of tears in Mauritius. Thanks to Günther, the bleak atoll had been a thorn in the side of the Mauritius government for some while.

By 1903, Günther was in his seventy-fourth year and was approaching retirement. He had done as much as he could for the Aldabra tortoises and believed that, whilst they might become extinct in the wild, at lease there would still be breeding populations on Curieuse and other Seychellois islands.

Nowadays we are used to campaigns that plead with us to save the whales, tigers, pandas and other animals, but Albert Günther was the first person in history to set out to try to save a species threatened with extinction. After 1900, ill health and old age considerably reduced his role in the study and future survival of the tortoise. Other scientists took over the mantle of his work but all of them acknowledged him as the font of wisdom on the subject.

Günther never lost touch with the plight of the tortoises and, when he died in 1914, he was still fretting over those on Aldabra. It is sad that he was never to know that his efforts were ultimately to be successful although, as we shall shortly see, perhaps not in the way that he would have imagined.

Galápagos or Bust

IN 1881, ALBERT Günther was making one of his casual morning patrols of the Natural History Museum's public display rooms (to where the British Museum's natural history collections had been moved the previous year) when he spotted an adolescent boy wildly enthusing on the virtues of natural history to his governess. A father himself, Günther was always keen to encourage the young and so he wandered over to see if he could be of help. As soon as the boy realised that he was being watched, he became quiet and withdrawn, hiding sheepishly behind his governess. Nonetheless Günther persevered and when the ageing scientist had explained who he was, the boy became animated again, chattering excitedly about what was obviously an obsession with wildlife and natural history.

Günther was quite taken aback by the youngster's vast store of knowledge. This strange child could name all the species of butterfly on display and frequently talked of the animals that he had in his own 'museum'. Half an hour later a bemused Günther made his excuses and left to return to his own researches, thoroughly mystified by the encounter. Although he did not know it at the time, the thirteen-year-old whom he had just been entertaining was heir to one of the most wealthy and powerful families in the world.

Walter Rothschild (whose full name was Lionel Walter Rothschild) was the eldest son of Baron Nathan Rothschild, head of the Jewish banking family whose extraordinary wealth and philanthropic behaviour had made them a household name.

Walter was born in 1868 and was brought up on his father's gigantic estate near Tring in Hertfordshire. As he was educated at home, isolated from the company of other children, it is perhaps not surprising that Walter proved to be a somewhat neurotic child. He was withdrawn, exceedingly shy and had a speech impediment, which meant that he found it difficult to control the sound level of his voice.

Uninterested in formal education, the young Walter Rothschild instead displayed a passion for the natural world, collecting insects, eggs, stuffed birds and almost any other animal that came his way. This obsession had become apparent in him very early. When only seven, Walter informed his parents that he wanted to own a museum. As if to prove the point he converted a garden shed into a mini natural history museum where his delicate collection of pinned butterflies and other natural treasures were placed on display. Walter's father

Walter Rothschild aged nine years. Even at this age he was already an expert naturalist.

severely disapproved of his son's peculiar behaviour, believing that he should be preparing himself for a career in the family's merchant bank. Walter, however, had no such plans and continued to be obsessed with the natural world.

The chance meeting with Albert Günther was to turn out to be highly fortuitous for both parties. Walter was a regular visitor to the Natural History Museum and Günther, as its keeper of zoology, now had a free rein in one of the largest (if not *the* largest) natural history collections in the world. In the coming weeks and months, Günther met Walter on several occasions, during which the boy would talk animatedly about the museum's collections, even daring to correct the labelling on some of the exhibits. A friendship of a sort was struck up between the fifty-one-year-old Günther and the thirteen-year-old Walter.

In 1882, starved of human company, Walter wrote his first letter to Günther. During the next eight years he was to write more than 320 others, sometimes at the rate of more than five a week and often stretching to eight pages or more. It is quite clear from these letters that Walter was pursuing with some vigour his ambition of creating his own natural history museum, gathering insects, birds and other specimens from across the world.

In July 1884, Günther was invited to visit Walter's Tring museum for himself. The invitation was gratefully accepted although the visit was less successful than either party might have hoped, after Baron Nathan Rothschild (Walter's father) made it quite clear to Günther that he disapproved of his son's fascination with wildlife. Although Walter was unaware of it for some years, the baron's warning was to ensure that Günther kept the boy at arm's length. This reticence only increased when, in later years, Walter's passion for collecting animals reached manic proportions.

In his memoirs, Günther was to write of his first visit to Tring:

[Walter] sought to cultivate my friendship which developed so far that I received invitations to Tring. There I met with a most friendly reception on the part of Lady Rothschild but less so from the father. I fancied that the latter did not look with satisfaction on the son's devotion to the study of Natural History. In this I was not mistaken and I determined not to assist the youngster in any of his big undertakings for forming a large collection (much as I rejoiced in them) but merely to maintain a very friendly intercourse with Walter and his family. Walter, however, never lost an opportunity of showing me personal attachment and I returned those feelings of hearty regard.

Günther seemed to believe that he might be regarded as a substitute father-figure for Walter although the power and wealth of the Rothschilds must have instilled a certain amount of wariness into the venerable old naturalist. Nonetheless, the two kept in close touch. As the years passed Nathan Rothschild accepted that his son's interests lay in the natural rather than the financial world and allowed him to study zoology at Cambridge University.

On his twenty-first birthday, Walter was given a parcel of land in Tring and the money to construct several buildings there. His dreams of owning a museum were becoming a reality and not before time: at that point Walter had already managed to amass a collection of some 38,000 pinned butterflies, 5,000 stuffed birds and around 3,000 other creatures. Walter, who had by then fallen out with almost everyone at the London Natural History Museum apart from Günther, boldly stated that his museum would house the greatest collection of wildlife, alive or dead, that the world had ever seen. This megalomaniac passion for collecting was to drive Walter during much of the rest of his life and would eventually have a great bearing on the world's giant tortoise population. In the meantime, the first live exhibits (four kangaroos and some emus) arrived at Tring.

In 1892 the Rothschild Tring Zoological Museum was opened to the public. In the park Walter had built a menagerie in which could be seen wolves, zebras, dingoes, cassowaries and other exotic beasts. Despite his desire to remain at a distance, Günther had provided much technical advice during the museum's establishment and had even advised his young protégé on its choice of curator. Tring was Walter's pride and joy. He devoted much of his time to travelling the world, looking for new and interesting specimens to populate it, writing long letters to Günther asking for his advice or, in some cases, gloating about purchases that, through his overt wealth, he could make but which the Natural History Museum, with its small government budget, could not.

Indeed, it was Walter's family wealth that underpinned the entire venture, for even though Tring attracted its fair share of visitors, the rate at which Walter spent his money on foreign collecting expeditions was prestigious. Between 1893 and 1908, Walter funded over 300 collecting expeditions that took in every continent and stretched from the Arctic Circle to the Falkland Islands.

Unfortunately, Walter Rothschild seems to have belonged to that class of Victorian naturalist whose interest was less in the beauty and the biology of animals than in owning them. Whereas some rich playboys take delight in collecting cars, houses and wives, Walter was determined to use his fortune to obtain for himself the largest and rarest animal specimens on earth. It was like stamp collecting but on a gigantic scale.

Walter's closeness to Günther meant that some of his mentor's interests and obsessions were bound to wear off on him. Early on in their friendship Günther introduced Walter to the marvels of the giant tortoise and the boy immediately took to them. As it had for Günther, Walter's interest developed into something of an all-consuming passion. Unlike Günther, however, Walter had money to indulge his obsession and, once the Tring zoo was up and running, also the space to house large numbers of tortoises.

Thus it was that, shortly after the opening of his zoo, Walter decided that he needed a giant tortoise for himself. Naturally, he wanted the largest one possible but the biggest known specimen to date had been the one that Günther had spent a small fortune shipping from Mauritius to the London Zoo. Walter had to make do with the second-largest known specimen (a 250-kilogram monster also resident in Mauritius), which was duly shipped to Tring in August 1893.

The fact that giant tortoises were large, difficult to obtain and nearly extinct fired Walter's desire to own more of them and during the next few years he set about trying to gather every available giant tortoise specimen − dead or alive. Within a short time tortoises were being shipped from locations worldwide to Rothschild's Tring museum or, in some cases, to his French chateau, located just outside Paris.

Günther's many public battles on behalf of the giant tortoise had not gone unnoticed by Walter and after struggling free of his parents' control he turned his attention to their plight. Günther's belief that the tortoises could not be saved by leaving them in the wild seems to have rubbed off on Walter, who, in 1896, began a massive and world-wide collecting expedition, the objective of which was to collect 'every tortoise they could lay hands on and also to collect all skulls, bones and carapaces of dead tortoises on all islands'. The first islands to receive this attention were the Galápagos.

True to his nature, Walter began planning in secret for his great Galápagos expedition, ensuring that any news of it was purposefully kept hidden from the entire academic community, including Albert Günther.

Walter was never one to rough it and rarely went on his own collecting expeditions (indeed, there were so many that he would never have had the time). He therefore employed Frank Webster, a professional animal collector, to organise this one for him. It took Webster (who, in the end, did not join the expedition) only a few weeks to make the arrangements. 'Your proposals,' he wrote, 'are satisfactory and your wishes shall be carried out.'

The chief naturalist for the expedition was Dr Charles Harris, a gifted taxidermist who could skin 'fifty birds a day'. Dr Harris was obviously concerned by Walter's instructions to collect everything in sight, as in February 1897 he wrote to Günther to seek approval for the plan.

Günther, who must have been a mite annoyed at receiving news of Walter's Galápagos expedition second-hand, nonetheless endorsed it, telling the naturalist to remove any and every tortoise that they encountered. This, like his plan to evacuate the Aldabra tortoises, was in the belief that to leave them put would surely spell their extinction. Later, in a letter to Walter, he explained his thinking:

My chief reason for telling Dr Harris to bring away every tortoise they saw big or little, alive or dead, was that the Orchilla moss hunters had already reduced them by more than half since Dr. Bauer was there in 1892, and they would have eaten them all in two or three years more; and I wanted to save them for science. Bauer found over 100 [tortoises] on Duncan [Pinzón] Island in 1892 and Harris carried off all those that were left, which was 29, so some 80 tortoises had been eaten in five years.

Under the command of Captain Samuel Robinson, the expedition left by boat from New York on 29 March 1897 and made good time to Panama, where the first round of general specimen collecting was to take place. They had only been ashore a few days before things started to go seriously wrong. First was the discovery that the third man in command, Otis Bullock, had a serious drink problem and that his drunk and disorderly behaviour was endangering the expedition by upsetting local people. Bullock was dismissed and ordered home but, shortly after arriving in New York, he died of yellow fever. News of his death reached the papers and, because of the Rothschild connection, accusations were made that the expedition had been ill conceived, under-prepared and badly run, none of which was true. Unfortunately, Bullock's death was just the beginning.

Within days, Captain Robinson also developed yellow fever and, despite fighting for his life with all his strength, he died on 9 April. Harris was distraught at the thought of the wife and three children that Robinson had left behind. He was also aware that his expedition was falling apart.

In fact, it now consisted of only three people: Charles Harris, James Cornell and George Nelson. Shortly after Robinson's death, Nelson deserted the party, fleeing back to New York with some money sent to him by his parents. Harris was outraged, writing to Walter: 'If I was his father I would not send him money to get home but send two parties [with] a kettle of tar or a bag of feathers with instructions to apply them liberally. Such rank cowardice is beyond my conception.'

With only two of them left and the fact that most of Walter's money had been spent on doctors' and undertakers' fees, Harris and Cornell reluctantly abandoned the expedition and bought themselves passage on a boat to San Francisco. As a final insult, on the boat both men were taken seriously ill with yellow fever. Harris recovered but Cornell died and was buried at sea on 2 May. Walter's great Galápagos expedition had ended in utter disaster before it had even begun.

It was a weakened and desolate Harris who stepped ashore in San Francisco on 10 May. He was the expedition's sole survivor but, despite everything, was still determined to push ahead to the Galápagos. 'Don't let this fall through if you can help it,' he wrote to

Walter. '$1000 more I think will carry the expedition from San Francisco [to the Galápagos] and return in good shape. I will carry this through and make it pay in the end.'

Walter relented and sent the cash. After much prevaricating, Harris managed to charter the *Lila and Mattie*, a schooner under the command of one Captain Linbridge. Shortly before leaving port, Harris wrote to Webster, apologising that 'the trip will cost nearly twice as much as first contemplated but at the prices you say the [animals are] worth I will bring back $20,000 worth for you unless the vessel sinks and then we all have something to lose.' He signed the letter, 'C.M. Harris (Galápagos or Bust).'

The *Lila and Mattie* set sail for the Galápagos on 21 June, arriving at the islands just over a month later.

Walter's initial plan was to collect every tortoise and bird from every Galápagos island but this proved to be typically overambitious as neither he nor Dr Harris had taken account of the terrain. As it had for every other visitor to the Galápagos, the sight of the harsh, blocky, mountainous volcanic islands with their relentless scrubby vegetation unsettled the collecting party, none of whom had ever visited them before.

The first few islands reached by the expedition were small and desolate, and historically had never been known to have populations of tortoises. Next they arrived at Pinzón Island, whose landscape was so forbidding that it took over a day for them to find a secure anchorage.

When, on 4 September, the men actually landed ashore the island proved to be every bit as inhospitable as it had looked from the ship. No water could be found, ruling out the possibility of setting up a suitable campsite. The party would have to operate using the boat as a base, complicating matters greatly.

More worrying still was the lack of any tortoises. A century's worth of visiting whale ships had taken all those that lived close to the seashore and the only remaining animals were to be found high in the mountains where the deep volcanic crevices and craters rendered them inaccessible to most people. Harris and his company realised that they were in for a far tougher time than they had at first anticipated, and so it proved to be.

One of Harris's shipmates, Frederick Drowne, kept a diary of his time in the Galápagos. His entries for their visit to Pinzón Island tell us just how bad conditions were, both for the men and the tortoises:

Sunday Sept. 5, 1897. After a long walk I arrived at the edge of the crater at about 11 a.m. Harris was already inside. We climbed down the side, I should say 250 feet [76 metres], and reached the bottom, which was level and covered all around with thick bushes on the border . . . Soon after reaching the bottom I heard Harris calling out that he had caught a tortoise. Hull and myself got there as soon as possible, and we tied the tortoise up. The grass was full of tortoise trails, and we set out in search of others. Harris found two more, and Hull and myself each two. We turned them all over, and weighted them down with heavy rocks. After fixing the last one, we revisited the first and found it loose. This made it necessary to revisit the others, which we did, finding that they had all got loose. We weighted them down again with more and heavier rocks, and returned to the starting-place. Some of the tortoises which we found feeding were eating the blossoms from a creeping vine, rising upon their forelegs and stretching their necks out to full extent. The odour from them reminded me very much of that from an elephant. After tramping about so much and lifting so many heavy rocks, we were very tired, but had to brace up and climb out of the crater, and walk to the shore over a long distance of broken rock. The crater was quite three-quarters of a mile [1.2 kilometres] in diameter, with a very flat bottom, surrounded by a high wall or embankment, making it resemble greatly pictures of the old Roman amphitheatres. Arrived on board at 6:30, very tired and very thirsty.

Tuesday Sept. 7. Another hard day's work. Got up at 4:45 a.m. and started to heave up anchor. Sailed over to Duncan [Pinzón] Island. Had breakfast at 6:30, and went ashore soon after, starting immediately up to the crater, with poles, ropes, etc., to get the tortoises out. Managed to recover our tortoises of last Sunday, some of which had got away. Found one dead, a rock having fallen on his neck during his struggles and shut off his wind. Found one more, making a total of eight. The work of making them fast lasted till about 2 o'clock, when we started for the store with a tortoise strung on a pole between each two men, one of the sailors and myself taking one. It was very hard getting them up the side of the crater, walking being so rough and thorns so plentiful. But this was nothing to be compared with going down on the other side, which was very steep and terrible walking. The sailor had on a pair of wooden clogs, which soon began to chafe his feet. After a long time spent in tumbling over lava blocks, tearing through thorn bushes and other such pleasantries, we reached a point as near the shore as we could, tied the creatures up securely, and left them. Now came a long walk before we could get to the skiff. We were all so tired, having had nothing to eat since breakfast, that the distance seemed terribly long. It was a rough road, up and down, over broken lava and through thorns. Reached the skiff about 6 p.m., every one being well tired out. A good drink of wine and water was served with the lunch that was in the boat. We got aboard the schooner a little later. This was the hardest day's work thus far, with the possible exception of last Sunday's. The trip was very hard on the tortoises also, and they acted as if 'played out'. Two of them being set down close together got their poles somewhat tangled up, and by the way they opened their mouths at each other it looked as if they were going to have a fight.

In total the party struggled to remove twenty-nine live (and two dead) tortoises from Pinzón Island, believing (erroneously) in doing so that they had emptied the island of its entire tortoise population. As this was one of the expedition's first ports of call, their response to its harsh landscape seems to have shaped the collecting strategy for the rest of the voyage.

A few days later, the expedition reached Santa Cruz Island where they met Thomas Levick, an Englishman who was travelling around the Galápagos with a small party of men. Mr Levick told Harris that all the islands were now devoid of tortoises, apart from Isabela where they were still relatively common. The rest, said Levick, had been killed off by the rats, dogs, cats and goats that had been introduced into the islands earlier in the century. Although this was not true, Harris and his crew chose to believe Levick. Despite going on to visit another ten islands, the party only collected another thirty-one tortoises from the lowland area around south-eastern Isabela Island. It is clear that Dr Harris was not prepared to risk the safety of his crew, searching for tortoises from the highland interior of the other islands where, as we now know, there were tortoises to be found.

The half-hearted tortoise collecting did not, alas, extend to the other animals of the Galápagos. True to his word, Dr Harris set about shooting and preserving anything that was foolish enough to move within eyesight of the men. Several thousand birds were shot as well as an unspecified number of reptiles and mammals. The carnage was terrible. Not that Walter Rothschild minded. He was delighted and wrote to Günther, boasting that 'if all's well [Harris] hopes to land at San Francisco 60 live tortoises, 20 dead ones, 200 iguanas and lizards, 100 mammals and 5,000 to 6,000 bird skins besides what he calls side dishes by which he means shells, insects and sea products. I think 60 living Galápagos tortoises will make people stare.'

Finally, after visiting the remote Genovesa Island and with the weather closing in, Harris ordered the captain to leave the Galápagos and to set sail for San Francisco, arriving there on 8 February 1898.

By now Dr Harris had become irritated by the whole venture and was less than pleased to receive some stern communications from his paymaster in Tring. Walter was annoyed that the party had missed out the northern part of Isabela Island and that they had refused to travel on to the Cocos Islands (because of a yellow fever epidemic in the area) as he had requested. Harris, who had expended the last of his energy in San Francisco sorting the dead specimens into sixty crates that were to be freighted to Tring, was ordered by Walter to act as chaperone to the sixty live tortoises. This was the final straw for Harris who promptly announced that he had had enough, was quitting San Francisco and was going to travel east to New York.

Walter gave strict instructions that Harris was to do no such thing and that he was to see to the tortoises' every requirement until they could be sent to England. Harris relented and spent several miserable weeks tending the beasts, hiring a heated greenhouse for them to live in, obtaining bananas for them to eat and shovelling fresh sand whenever they needed it. Eventually the animals were crated up and started their long journey to meet their new owner. In July 1898, Walter greeted them with open arms, informing Günther, 'I now have fifty-five live Galápagos tortoises to play with.'

Together with his previous acquisitions, Walter now had the world's largest private collection of giant tortoises, all of which were housed in large wooden pens at Tring. On summer days he would

Walter Rothschild rides Rotumah, a Galápagos tortoise that he found
living in the grounds of an Australian lunatic asylum.

take great delight in getting astride some of his larger specimens and riding them slowly round the gardens.

The Harris expedition had cost Walter well over $3,000, a considerable sum, but this was only the start of his tortoise-collecting mania. Over the ensuing years he spent thousands of dollars buying up dead and living giant tortoise specimens from around the world. By the early 1900s there were over 144 live animals at Tring and many, many more dead specimens (by 1896 the dead already numbered 209). These were Walter's pride and joy, and with each acquisition came a flurry of letters from Walter to Günther, expounding the virtues of the new specimens.

Some of the stories were indeed extraordinary. In 1898, Walter found 'Rotumah', a massive Galápagos tortoise that had been left on the Marquesas Islands by Captain Porter in 1813 but which had subsequently been moved to the grounds of a lunatic asylum in Sydney, Australia. After much bartering Walter purchased Rotumah and had him shipped to Tring where, two years later, he died of 'sexual over-excitation'. Another three batches of Galápagos tortoises followed, plus many individual specimens from Aldabra. The Natural History Museum's tortoise collection looked puny in comparison. All this collecting may have satisfied Walter's schoolboy mind but it was the tortoises themselves that were paying the price. By Walter's own admission, no captive giant tortoise had ever survived in England for more than fifteen years. For most it was considerably less than this.

Tortoises that had lived for several decades in the wild would last only a few months in the cold European climate. A tortoise from the Sandwich Islands, which was over a hundred years old, was dead within months of its arrival, and the same was true for a multitude of others brought to Europe. Given that any tortoise that was transported into the northern hemisphere was doomed to die prematurely, one must question the sanity of the actions of Rothschild and Günther, both of whom had backed various plans to uproot tortoises from their native islands in order to 'protect' them.

That Günther, a conservationist at heart, approved of Walter's 1897 'rescue plan' for Galápagos tortoises is still a puzzle. In the course of that expedition over 11,000 animals were killed or kidnapped, and Walter's boast of having 'emptied Pinzón Island of tortoises' seems to be at odds with the idea of conservation.

Through the eyes of the modern-day naturalist, whose general belief is that removing animals from their native environment is a last resort, these expeditions could be seen as devastating. It is perhaps possible to see that transplanting some of the tortoises might have saved them in the short term and allowed their reintroduction into the wild at a later date, but quite how Günther and Rothschild believed that the killing and skinning of several thousand birds was going to help matters is beyond imagination. Again, it is the difference in the mentality of Victorian scientists and our own. No doubt in one hundred years' time, scientists will respond to our zoos and captive breeding programmes with horror. Such is the wisdom of hindsight.

The Errant Playboy

IT WAS PERHAPS inevitable that Walter Rothschild's ambitions would stretch further than just having the greatest natural history museum on earth. He wanted to be taken seriously as an academic too.

It is clear that from early on Walter had a great distrust of many academics. It is also quite probable that because of his mildly paranoiac nature, he may have viewed most established scientists as some kind of threat or at least disapproving of his amateur efforts. Günther, as one of the few scientists allowed into the inner sanctum of Walter's life, was frequently called on to intervene in the various spats that erupted between Walter and staff at the London Natural History Museum.

For example, Walter once wrote to Günther asking him 'to tell Boulanger [Günther's successor at the museum] from me that I consider it a mean and ungentlemanly trick to name a tortoise of mine without permission. In fact, you can tell him that I shall send all my Galápagos reptiles to Frankfurt and not to him as I promised last year. I have a good deal of influence among our people, I shall not hesitate to show men of Mr Boulanger's type that they make a mistake in being too bumptious.'

Günther managed to pour oil on the troubled waters, calming Walter down and getting Boulanger to apologise. As an act of reconciliation Rothschild eventually named the specimen *Testudo boulangeri* although as the name does not seem to have made it into print, it is thus invalid.

In the academic world the major measure of success comes from the number of printed papers that an individual scientist can attach to his or her name. From an early age Walter had a strong desire to get his writings into print but both his youth and lack of formal scientific training worked against him. From the outset he sought Günther's help in this matter. Although the old gentleman was willing to read Walter's draft manuscripts, he refused point-blank to put his own name to them or to use his influence to get them accepted by learned journals.

Walter, impatient with journal editors, was frequently rude to them when they questioned aspects of his work, writing them long ranting letters. Günther, who was usually left to pick up the pieces, was often treated as Walter's factotum. 'If it is not bothering you too much,' wrote Walter on one occasion, 'I should be most grateful if you would re-write my article.'

Eventually Walter solved the problem of getting his work published in the only way that he knew how – by throwing money at it. At twenty-five, he founded *Novitates Zoologicae*, his own personal scientific journal, of which, naturally enough, he was editor. *Novitates Zoologicae* was to publish the majority of his writings with, more often than not, the patient Günther acting as his proof-reader. From the moment that Walter started to accumulate giant tortoises at Tring he put pressure on Günther to join with him in writing a massive and definitive work on the world's giant tortoises. This pressure continued for many years. Günther, whose 1877 monograph was still the standard work on the subject, was initially keen but was deterred by Walter's father, who made it apparent that he did not approve of the huge expense involved in producing these enormous monographs. Although Walter produced vast numbers of papers on other subjects, he seems to have been wary of standing on Günther's toes and during the old man's lifetime only wrote a couple of short papers on the giant tortoises.

In fact, Günther was to update his own landmark 1877 work in a brief monograph in 1898. It was only in 1915, the year after Günther's death, that Walter produced his own monograph on the subject. It was almost as though Walter was waiting for his mentor to pass on before he felt brave enough to put into print his own ideas about the tortoises.

One reason for Günther's reluctance to collaborate in print with Walter was the young playboy's immature personality. As a boy Walter had been socially crippled by shyness, and in adulthood he was still emotionally handicapped, as well as having an unfortunate speech impediment. He found it difficult to make social connections and never married. He did, however, have a string of affairs and, at one point, kept two mistresses in London. It was to be one such sexual indiscretion that was ultimately to bring about his downfall.

The Rothschilds were rich beyond comprehension, so that the thousands upon thousands of dollars that Walter threw at his museum and collecting expeditions were easily absorbed by his fortune. In the late 1890s Walter had an affair with a married woman who was well connected. Very little is known about this liaison but, some time around 1892, the 'peeress' (as this woman became known) started to blackmail him, threatening to reveal the truth of their affair to his family. Walter was terrified of bringing shame to the Rothschild name and agreed to pay up but the blackmailers (the peeress' husband was apparently also involved) kept coming back for more, demanding ever-increasing amounts.

By 1908 the sums being extorted from Walter were so exorbitant that they began to affect the rest of his estate. In an effort to compensate, he began to cut back on his collecting expeditions but, as time went on and the blackmail continued, he was forced to sell some of his precious wildlife collections. Nonetheless, during the whole time that this was happening Walter did not breathe a word of the blackmail to anybody just in case word should get back to people whom he respected and who knew nothing of his womanising.

It was early in 1914 that tragedy occurred. The previous autumn had seen Günther in hospital with stomach pains, which necessitated an operation. Günther, who was by then in his eighty-fifth year, never fully recovered from the trauma and struggled on for a few months before finally dying on 1 February. Walter was devastated and wrote to Günther's son: 'The inevitable kindness shown me by Dr. Günther ever since my early childhood made me feel, when I learned the terrible news, almost as if I had lost a parent. In everything connected with my love for natural history Dr. Günther took such a vivid interest and used to help me in so many ways that his loss will cause an irreparable void in my life.'

Günther's part in the saga of the giant tortoises was over but his contribution would not be forgotten. He was the first person to wake up to their plight and he may well have given those on Aldabra enough breathing space to be able to survive into the twentieth century, when their numbers began to recover. He had also been the first scientist to tackle the problem of the number of tortoise species worldwide. Before Günther, only a handful had been identified. He was the first to realise just how diverse the tortoises were and, consequently, how much damage had been done to their numbers.

Günther's death marked a watershed in Walter's life. Shortly afterwards Walter's own father passed on, leaving his son at the head of a branch of one of the richest families in the world. Walter (now the new Lord Rothschild) had for some time been involved in British politics and the baron's death greatly increased his duties. This, and other formal obligations, began to eat away at his free time. The blackmailing peeress was also still active, sapping him of money and forcing him to cut back on any non-essential expenditure. It was the start of a long, slow, downhill slide and, whilst his interest in natural history never died, it had more and more to play second fiddle.

By 1931 the blackmailers' cash drain was such that Walter was forced to sell his cherished and magnificent collection of stuffed birds to the American Museum of Natural History for $225,000. *The Times* was mystified, describing the sale as a 'loss to British ornithology', little knowing the real reason behind Walter's actions.

In his final years Walter's decline was rapid as he succumbed not only to the financial pressure of his blackmailers but also to ill health brought on by his passion for rich food and fine wine. In the end it was not his lifestyle that was to claim his life but cancer of the spine from which he died in 1937.

It is widely acknowledged that the death of Walter Rothschild marked the end of the era of the amateur Victorian scientist, of which he was a spectacular if somewhat eccentric example. At the last minute he decided to donate his entire collection of animals to the London Natural History Museum instead of breaking it up, as he had originally intended. The museum gratefully accepted the gift and the Tring collection remains an important part of its set-up.

On cataloguing their new acquisition, the Natural History Museum discovered that the little boy who, aged seven, had filled a garden shed with butterflies had in the space of fifty years managed to accumulate 2.5 million butterflies, 300,000 bird skins, 200,000 bird eggs and 30,000 books. He was also directly or indirectly responsible for naming 5,000 new species and had overseen the publication of 1,200 books and papers. A staggering achievement by anybody's measurement.

Walter's once unparalleled collection of living giant tortoises had declined greatly as the British climate gradually claimed the lives of his charges. As an economy measure, those that lived on had been donated to the London Zoo. On hearing of Walter's death, the Zoo felt obliged to offer to return the tortoises to his niece Miriam. She refused the gesture.

In retrospect Walter Rothschild and Albert Günther had a mixed effect on the survival of the giant tortoise. They certainly helped in our scientific understanding of these strange creatures but their habit of removing them from the wild probably did more harm than good. Nonetheless, thanks to Albert and Walter the world was at least aware of the fragile situation of the tortoises. If they were to be saved, it was now the task of others with a better understanding of conservation and ecology to take their concerns and turn them into real action.

PART FIVE

Pets

The Well-Travelled Tortoises

THERE IS ONE aspect of tortoise biology that has always fascinated mankind. As long ago as 1810, when animal matters were not high on the list of many people's priorities, the prestigious *Times* newspaper of London ran a piece on a tortoise living in the gardens of the Bishop's Palace at Peterborough. This tortoise, reported *The Times* with some amazement, 'is ascertained to have been there for 200 years and upwards!' If this is true (and that is by no means certain), the bishop's tortoise was alive and well several years before Shakespeare wrote his final play.

The remarkable longevity of tortoises, especially of the giant tortoise, makes them unique among the world's animals. Whereas the average human may expect to own and outlive several pet cats or dogs during his or her lifetime, a captive giant tortoise may in turn be expected to outlive several of its human keepers.

Nobody is really sure how old a tortoise can get. Most of the aged specimens alive today were taken from the wild and so their exact year of hatching is not known. Those born in captivity are still comparatively young, being less than a century old at most. The current record is claimed by a Galápagos tortoise living in a zoo in Queensland, Australia, that is said to have hatched around 1830, which, at the time of writing, makes it approximately 173 years of age. (There are, as we shall see, some doubts about the provenance and age of this tortoise.)

Other historical records, including that of the bishop's tortoise cited above, claim tortoise ages of 200 years or more but proving (or, indeed, disproving) these claims is nigh on impossible. It can be said with reasonable certainty that particular tortoises have lived to be one hundred or more, and it would not be surprising to find that they are capable of surviving to a much greater age than this.

This extraordinary longevity not only makes the giant tortoise the longest-living vertebrate species on earth but also means that individual animals have been alive for so long that they have accumulated quite a personal history. Many have passed through the hands of several owners and have been shipped around the world.

A good example is the giant tortoise that belonged to the adventuring mariner Captain James Cook, which was probably picked up by him either in the Indian Ocean or second-hand in New Zealand. In 1777, Cook gave the tortoise to the king of Tonga as a present. The king was delighted, named the animal Tu'lmalila and gave it pride of place in the gardens of the royal palace. It even had its own personal keepers. In time the people of Tonga grew to believe that the tortoise itself embodied some of the supernatural powers of a chief and revered the animal. In May 1966, after long decades of pampering, the tortoise eventually succumbed to the laws of nature and died. Assuming that the historical records are correct, and they seem to be, this animal was approaching the magical 200-year mark. The stuffed corpse was later donated to Auckland Museum in New Zealand.

Many other tortoises were uprooted from their homes on the Galápagos and Indian Ocean islands and shunted from pillar to post before finally coming to rest in places that could be thousands of kilometres from their native islands. Usually the origins of these animals faded away with time, their original owners having long since died. It is only in the last hundred years or so that efforts have been made to try to track down the provenance of some of these dispersed tortoises. This process has thrown up some surprising results. The folklore that has come to surround some of these animals is amazing and this is nowhere better illustrated than in the curious case of Mr Darwin's tortoise.

Darwin's Missing Tortoise

IN JULY 1994 a newspaper in the Australian city of Brisbane published a reader's letter about the desperate plight of the Galápagos
tortoise in the modern world. Its conclusion was that humans were
responsible for their near-total annihilation. Dozens of such letters
are written and published around the globe every day, most without
raising anything more than a murmured comment across the breakfast table, but this one was different. It would set in motion a train of
events that eventually offered a direct link between the modern age
and the early life of Charles Darwin.

By chance the letter was read by Ed Loveday, an elderly former historian who was living out his retirement on the outskirts of Brisbane.
The mention of tortoises jogged a distant memory from Ed's past,
which, for some reason, he felt needed sharing with his fellow citizens. He cogitated for a week or so and then put pen to paper, formulating his own letter to the newspaper, which published it.

'Tortoise Recalled'

The sad story of 'George', headed 'Last of his kind' in the *Sunday Mail*
of July 6th, reminded me there were once three Galápagos tortoises
living in the old Brisbane Botanical Gardens.

In my time, from about 1922 onwards, there were only two still
living. I was told they were brought to Brisbane by Captain Wickham,
the Government Resident at Moreton Bay around the middle of last
century. Wickham had accompanied Charles Darwin on the *Beagle* on
his research voyage around the world and spent some time at the
Galápagos islands gathering scientific material for Darwin's classic
work *Origin of the Species* [*sic*]. It is quite probable that Wickham took
the three specimens from there and later installed them in the
Botanical Gardens where I saw them several times.

Eventually all died, the last one fairly recently. Sadly, they did not reproduce; I never heard of this anyway. Perhaps they were too old when coming to Brisbane, or were of the same sex or they enjoyed the lush Botanical Gardens conditions and did not bother about that. Certainly they were early residents of Old Brisbane Town. They were very long lived. Perhaps others of your readers could add to this reminiscence.

On the face of it Ed's letter offered a remarkable story, claiming that three giant tortoises, once resident in Brisbane's Botanical Gardens, had been brought there from the Galápagos Islands via Charles Darwin's ship the *Beagle*. Unfortunately, even though Ed's story was fascinating, it did not appear to have much of a happy ending: the tortoises were now all dead, severing any links to Darwin. Or so Ed thought.

Several hundred kilometres away, in the city of Canberra, was Dr Scott Thomson, a reptile specialist who had a long-standing fascination with Galápagos tortoises. In recent years Thomson had been trying to trace the whereabouts and origins of every single one of these giant beasts to have been imported into Australia and New Zealand.

As with all historical research of this nature, Thomson had had many successes but there were still a few failures. Uncovering the origin of one giant tortoise in particular had proved to be especially problematic.

The tortoise concerned is known as Harriet, a large adult female weighing in at some 180 kilograms, which is big, even for these animals. For several years Harriet had resided in the Australia Zoo, a large complex of buildings near Queensland's famous Sunshine Coast. The Australia Zoo's central attraction is its extensive collection of large and dangerous crocodiles, which the staff take delight in hand-feeding in front of crowds of visitors. Harriet, a slow ponderous animal that spends most of her days hiding in the shadows, could not exactly be called a crowd-puller. Her large and well-tended pen is located in a remote part of the zoo, near the kangaroos and koalas but some distance from the snapping crocs. In relation to the crocs and koalas, Harriet gets few visitors.

When Scott Thomson first heard about Harriet in 1992, the news of the existence of such a large and therefore probably aged giant

Harriet the tortoise is listed as the world's oldest living animal but her origin is far from straightforward.

tortoise immediately caught his attention. He contacted the Australia Zoo's owner, Steve Irwin, a former government 'crocodile hunter' who is now famous for his television appearances and on-screen, crocodile-wrestling antics.

Irwin's obsession with matters of a reptilian nature is legendary and he was delighted that somebody had taken an interest in Harriet. Most people who approached him were only interested in the zoo's snakes and crocodiles; Thomson's tortoise project made a refreshing change. So, some months later, Thomson made the journey from Canberra to Queensland to meet both Irwin and Harriet.

Irwin confessed that he knew very little about Harriet, other than that she was a Galápagos tortoise and that she was quite possibly over one hundred years old. After seeing Harriet, Thomson was quickly able to identify her as being a Galápagos tortoise. He felt that she was probably *Geochelone nigra porteri*, which means that she would initially have been taken from Santa Cruz Island. For a scientist of Thomson's calibre, identifying Harriet's species was the easy part. Now he had to fulfil the second part of his research project: to trace Harriet's life history.

Anybody who has ever tried to do this with their own family history will testify as to what a laborious and frustrating process it is. Humans are relatively good at leaving written records of their existence, and documentary evidence such as birth certificates, census records and electoral registers can all be used to put together the life history of an ancestor. Giant tortoises, on the other hand, are not nearly so likely to have left their mark on society, making their life histories a great deal more difficult to uncover.

An adult Galápagos tortoise, such as Harriet, could have lived for a century or more, spanning the working lifetime of three or more human owners. As such, word of mouth alone would not be enough to define Harriet's origins. The person who originally took her from the Galápagos had most likely been dead for several decades. This was going to involve some serious and time-consuming historical research.

By good fortune, Thomson had already had a degree of success in researching other Australian Galápagos tortoises. He had tracked down two individuals that had been brought into Australia in the 1860s and had then spent their lives in the grounds of a Sydney mental institution before being shipped to England by Walter Rothschild. He had also managed to source the origins and whereabouts of a Galápagos tortoise imported into Sydney in 1870, as well as six more that had arrived in the 1930s. With such a good track record, there seemed to be no reason why Thomson should not be able to trace Harriet's origins as well. The starting point was, naturally, her current owner, Steve Irwin. Did he know much about her?

When Thomson first made contact, Steve Irwin had only been in charge of the Australia Zoo for a few months, as it had previously belonged to his parents, Bob and Lyn Irwin. After years of working for Queensland's rogue crocodile relocation programme, Steve had returned home in 1991 to take over the family business that, as a child, he had helped to build into a successful tourist attraction.

Harriet was at the zoo on his arrival but a quick glance through the paperwork revealed that, despite her age, she was not one of its long-term residents. In fact, she had only arrived just a few years earlier, in 1987. Her previous owner had been Fleay's Wildlife and Fauna Park, another zoo located on Australia's Gold Coast. Unable to get any further with the Australia Zoo, Thomson switched his attention to Harriet's previous owner.

This turned out to be one David Fleay who seems, by a twist of irony, to have shared a great many character traits with the current owner, Steve Irwin. Fleay, a native Australian, was born in 1907 in the small town of Ballarat, near the city of Melbourne. The young Fleay's interest in wildlife was instantaneous and all consuming. By 1931 he had a degree in zoology but was back in his home town of Ballarat, working as a teacher in a local school.

Like many teachers, he used his spare time to indulge in his passion for mother nature. However, by then he had become selective and only had eyes for native Australian fauna, on which he was rapidly becoming an expert.

In 1933 he travelled to Tasmania where he made a pilgrimage to Hobart Zoo, which, at that time, contained a single specimen of the Tasmanian tiger or thylacine. The thylacine, a most extraordinary animal, looked like a large but slightly misshapen dog with a series of distinctive black stripes running across its back, from which it gets its common name. Any similarity between the thylacine and the dog is purely coincidental; the former was a marsupial mammal (like the kangaroo) while the latter is a placental mammal (like us), so the thylacine has more in common with the kangaroo and the koala than the dog.

Thylacines were once a common sight in Tasmania but the poor beasts were mercilessly hunted for sport and for their pelts, so that by the opening decade of the twentieth century there were practically none at all left in the wild. Only a few individuals survived in captivity.

During his 1933 visit to Hobart, Fleay became the last person ever to have taken a photograph of a Tasmanian tiger. The animal itself rewarded him by waiting until his back was turned and then biting him hard on the buttocks. The scars, which remained with him until his death, were apparently a talking point at social gatherings. Despite his wounds, he immediately realised that the Tasmanian tiger was teetering on the edge of extinction and was inspired to try to save the species. He pleaded with those zoos that had thylacines to co-operate with one another and begin a breeding programme. They refused point-blank. Fleay had to watch as one of his most treasured native animals travelled the inexorable path towards extinction. The Hobart thylacine that had savaged Fleay's rear died in the autumn of 1936, terminating the lineage of a distinctive and completely unique Australian species.

As Fleay's reputation as a naturalist grew, so his home town of Ballarat became too confining. In 1934 the Melbourne Zoo founded a section that contained exclusively native Australian animals. When they needed someone to design and direct it, there was really only one choice. Fleay quit his teaching job and left Ballarat for Melbourne.

When it came to designing the zoo, Fleay's outlook proved to be some decades ahead of its time. He did not just want to keep his animals in cages for visitors to stare at; he wanted them to breed and increase their numbers so that they could be released back into the wild. Many native Australian species had come under pressure from farming and from competition with imported species such as cats, rabbits, rats and dogs. Rather than just building a zoo, Fleay was constructing what would today be called a 'conservation centre'. Clearly the memory of the thylacines' extinction still weighed heavily on his mind. The Melbourne Zoo never quite saw the point of encouraging their exhibits to breed when there were plenty more in the wild that could simply be captured. Nonetheless, Fleay managed to get koalas, emus and other Australian animals to reproduce successfully. However, there were frequent disagreements with the management. The arguments grew more intense, culminating in his being sacked in 1937 after he insisted on feeding his captive native birds on grubs and worms of the kind that they would have eaten in the wild.

Fleay accepted a position at a zoo in nearby Healesville where his breeding programme began in earnest and was quickly (and controversially) expanded to include highly venomous snake species, of which there are quite a few in Australia. Again, there seems to have been a certain amount of tension between the Healesville management and Fleay, which resulted in his taking some of his star animals, the platypuses, to be exhibited in America for a few months. On returning he found that he had been sacked in his absence although a public outcry forced the zoo to re-employ him.

It was clear that Fleay was never going to be comfortable working under somebody else's authority. For several years he had been nurturing at his home a private collection of native Australian animals in order to carry out his own personal research into their lifestyles and breeding habits. In 1951, the state of Victoria passed a law that effectively made it impossible for him to continue to do this. The time had come for him to set up on his own and so he and his menagerie relocated themselves to Queensland.

Fleay's Wildlife and Fauna Park opened its doors to the public in 1952. It offered a great many attractions, including a 'taipanery' where he reared several specimens of the deadly taipan snake, milking them for their venom, and a 'platypussary', where he bred platypuses. Indeed, until 1998, Fleay was the only person to have successfully bred this animal. However, as popular as the platypuses, emus, koalas and snakes were, his collection of native species was relatively small and he knew that in order to attract the crowds he would need to expand his range of animals.

The first opportunity to do so came soon after the opening of his zoo. In 1958, he got word that a small wildlife park in Brisbane was shutting down and that the animals there, which included several koalas, kangaroos and wallabies, needed a home. David Fleay drove down to Brisbane to see them.

The wildlife park concerned was in fact right in the heart of Brisbane, in the city's large and diverse Botanical Gardens. These had initially been founded in the mid-Victorian era by the Queensland Acclimatisation Society, whose members imported exotic species of plant to see if they would successfully take root in the Australian soil and climate. One century later, there was a concerted effort to uproot these imported alien species, many of which had found Australia so congenial that they were now ousting entire elements of the native ecosystem. Although essentially concerned with plants, the Queensland Acclimatisation Society also made room for that most Victorian of fascinations, a zoological park. However, in 1958 there was no longer either the time or the money needed to keep it going and the decision was made to shut it down. Any animals that could not be sold would be humanely put down.

Fleay arrived at the Botanical Gardens with the intention of reclaiming only the Australian animals, but as he wandered past the cramped cages and pens, something caught his eye. Whilst all the other animals were enclosed or tethered, one lone individual was left to roam free about the gardens, keeping itself to itself. It was a large tortoise whose slow and deliberate movements meant that it was unable to escape, even if it wanted to.

Fleay knew that the tortoise could not be a native Australian animal, since there are no native giant tortoises on the continent, but the sheer size of the animal made it difficult to ignore. After watching the tortoise plodding about on the grass for a few minutes, stopping occasionally to nibble at a flower, Fleay decided that it might form a good attraction at his new wildlife park. Snakes and crocodiles were all very well, but he was aware that he had a distinct lack of animals that people could get close to, or even touch, without the danger of being bitten, stung, kicked or scratched.

A quick enquiry revealed that the giant tortoise was an adult specimen known as 'Harry' that had been at the park as long as anybody there could remember. According to the park authorities, Harry was noted for being docile and tolerant, and in the summer months he would be brought out to give small children rides on his back. This was all Fleay needed to hear and that evening he set off home with an agreement to take several of the Botanical Gardens' native animals plus Harry the giant tortoise. Once they had arrived in his zoo, Fleay immediately discovered that Harry, the supposed male tortoise, was in fact a female; Harry, therefore, became Harriet.

Fleay, being curious by nature, tried himself to find out something of Harriet's past and, through one eyewitness account, managed to find evidence that she was in the Brisbane Botanical Gardens in the 1870s. This was to be the oldest known reference to Harriet that Scott Thomson would trace. Prior to this, her origins are a mystery.

During her time with Fleay, Harriet became a minor celebrity, featuring in his books and radio broadcasts and, being one of the few child-friendly animals in the park, she was popular with the crowds. When Fleay started to cut down on his commitments in the 1980s, Harriet was sold to Steve Irwin's zoo. Fleay himself died in 1993,

Harriet as she was in 1958, but could she really once have been
Charles Darwin's pet?

having received dozens of awards and accolades for his lifetime's work
with Australian (and other) animals.

At the end of his research, Scott Thomson was able to say with
certainty that Harriet had been in the Brisbane Botanical Gardens
from before 1870 until 1958, after which she had moved to Fleay's
zoo and then, in 1987, to Steve Irwin's zoo. Prior to 1870 most of
Brisbane's city records had been destroyed in a terrible flood, including the papers for the Botanical Gardens. As far as Thomson was concerned, the ultimate origin of Harriet would remain a mystery. Then
he was sent a copy of Ed Loveday's letter.

It was not difficult to accept the possibility that Harriet might be
one of the three giant tortoises referred to in Loveday's letter. If so, and
if Loveday's story about the tortoises' origins were true, then not only
was Harriet one of the oldest animals on earth, but she could also be
a living link between the modern world and Charles Darwin himself.

Thomson contacted Ed Loveday, who confirmed to him that what he had written in the letter was true. He added that there was another eyewitness who could corroborate that Harriet was in the Botanical Gardens during the 1920s. So far, both Ed's and Thomson's information tallied but there was still Loveday's central allegation that the Brisbane tortoises had originally come from Darwin's travels on the *Beagle*. Could Harriet really be a living legacy of Darwin's most famous voyage or was it simply too good to be true?

Digging into Harriet's Past

IT WAS BY pure chance that in the autumn of 2001 I had the pleasure of meeting Harriet the tortoise for myself. I was in Queensland on a ten-day work trip that took me from Brisbane to Fraser Island and back again. By good fortune Steve Irwin's Australia Zoo was only a short detour from my route and so, on a hot spring afternoon, I pulled into a car park that was already brimming with cars and tourist coaches.

Prior to my arrival I had arranged to meet a member of the zoo's staff in the hope of gaining a bit more information about their extraordinary tortoise. However, I had also chosen to arrive at the same time that the zoo's charismatic owner, Steve Irwin, was in the middle of making a feature film called, appropriately enough, *The Crocodile Hunter*. Consequently, the staff member I met was primed to talk about one thing and one thing only – Steve Irwin.

Each question I asked about Harriet was somehow turned around so that the answer concerned Mr Irwin's crocodile wrestling habits and, of course, his forthcoming film. When I suggested that we might pay a visit to Harriet, I was taken instead to see their collection of ferocious-looking crocodiles and told how, during a flood, Steve had wrangled a large escaped animal, measuring 2.75 metres, back into its pen but only after it had taken a chunk out of another keeper's leg.

At the end of my meeting I knew a great deal about crocodiles, deadly snakes and Steve Irwin but not much more about Harriet herself. However, in fairness, the Australia Zoo does not use its giant tortoise to pull in the crowds.

Having been left to my own devices I threaded my way back through the zoo in search of Harriet's enclosure. I finally found it tucked behind the kangaroos in a rather out-of-the-way location. Harriet's pen was large and the old girl herself was resting under some low trees in a far corner. Even from a distance it was possible to see that she was not only gargantuan in size, but that she was also getting on in years. Her eyes were slightly sunken and her movements slow and deliberate, even for a tortoise of her size. However, she looked happy enough and the decision to situate her pen away from the crowd-pulling crocodiles and snakes seemed to afford her the sort of peace and quiet that we should all expect to receive in our 172nd year (or thereabouts). I stopped some passing zoo-keepers and enquired about Harriet.

'She's a game girl for her age,' said a young male keeper. 'Always goes for a wander around her enclosure in the mornings and then sleeps it off in the afternoons.' Another confided to me that she was very fond of hibiscus flowers. Neither of them knew much about her past beyond what was written on the sign about her links with Darwin and the *Beagle*.

Time was pressing and, after a few minutes' gazing at the distant and now sleeping tortoise, I was forced to hit the road again, heading back to Brisbane. My brief visit to Harriet was to be the start of a long and complicated quest to try to understand her history.

Prior to seeing her in person, I had been made aware of Harriet's story through a newspaper article to commemorate her 170th birthday. As well as celebrating her longevity, the article also mentioned her connection to John Wickham, Charles Darwin and the *Beagle*. As a fan of Darwin, tortoises and unusual stories, the novelty of Harriet immediately seized my attention and I thought that I could see a good journalistic piece on the horizon.

A few days later, I had managed to find Scott Thomson's paper on Harriet's history, which outlined the connections between Darwin, Wickham, the Brisbane Botanical Gardens, Fleay and Irwin. On the face of it the case in favour of Harriet having been plucked from the Galápagos by Wickham seemed convincing. After all, she was definitely from the Galápagos and tests had confirmed that she was the right sort of age.

Nonetheless, I was also aware that, despite the word of Ed Loveday, there was still no solid evidence that Harriet had anything to do with Darwin. In fact, as we have seen, the first-hand historical trail for Harriet runs out in the 1920s when Ed Loveday remembers seeing her in the Botanical Gardens. If we accept David Fleay's word, that trail could stretch back to the 1870s but that still leaves a gap of forty-odd years between the *Beagle*'s return to England in 1836 and Harriet's appearance in Queensland.

My brief visit to the Australia Zoo increased my interest in Harriet and so I began to search for more proof of her history. Before leaving Australia I had managed to call at the national archives in Brisbane and Sydney in the hope of finding further references to Harriet, but in both cases drew a blank. As Scott Thomson had already noted, a great many of Brisbane's records were destroyed in a mighty flood that swept through the city in February 1893. Unfortunately, the destruction included documents relating to the Botanical Gardens, which may have contained early references to Harriet and the other animals in the zoo there.

On returning to England, I continued to research the subject and, since I had access to contemporary English records, it seemed more logical to focus on trying to find out what happened to the *Beagle* tortoises once they left the Galápagos Islands. This turned out to be much harder than I initially thought.

The published journals of both Darwin and Captain FitzRoy confirm that when the *Beagle* left the Galápagos she was laden with live tortoises as a food source. On 18 September 1835, while at San Cristóbal Island, the ship took on board eighteen adults, which Fitzroy notes were not 'exceeding eighty pounds' (thirty-six kilograms) in weight. Then, on 12 October another thirty large tortoises were taken on board, again at San Cristóbal Island. It is likely that all forty-eight of these large tortoises had been cooked and eaten long before the *Beagle* reached New Zealand two months later. Certainly none of their remains were still on the ship when she returned to the UK in October 1836.

However, whilst the larger tortoises may have been eaten, as previously mentioned some of the crew took juvenile tortoises on board to keep as pets. Captain FitzRoy retrospectively notes that 'On board the *Beagle* one small tortoise grew three-eighths of an inch [9.5mm], in length, in three months; and another grew two inches [5.08cm] in length in one year. Several were brought alive to England.' From this description it is not clear just how many juvenile tortoises were brought back and to whom they belonged but it seems likely that there were at least four specimens. We know this chiefly because of Darwin's conversion to the theory of evolution during 1837.

As we have already seen, he reflected on the words of the Galápagos governor who boasted that, when shown a tortoise, he could tell which island it had come from. Back in England, Darwin needed to examine some Galápagos tortoises in order to check whether each island did indeed have its own species. He later wrote that he had managed to source some Galápagos tortoises but that they were 'young ones and, probably owing to this cause, neither Mr Gray nor myself could find in them any specific differences'.

So, some time in early 1837, Darwin managed to gather together several young tortoises and took them to the British Museum in order to be examined by John Gray, the resident expert in reptiles. As we have already seen, the task was ultimately fruitless, the tortoises being far too young to show any real differences in their shell shape. Could one of these tortoise have been Harriet?

The precise fate of those young tortoises has previously attracted the attention of specialists. Adrian Desmond and James Moore, who wrote an authoritative biography of Darwin in the early 1990s, admit that their attempts to track down the *Beagle* tortoises were unsuccessful, while an ex-curator at Down House (Darwin's old residence) said that she had heard that the tortoises had been given to the Bishop of Llandaff. This last idea seems a little unlikely and I must confess that I have not been able to discover where and when this particular rumour started. Leaving the Bishop of Llandaff to one side, some of my recent research has shed some light on the eventual fate of at least some of the tortoises that Darwin and Gray examined.

It seems that the two naturalists probably had four tortoises in their possession that day. Two of these had belonged to Captain FitzRoy while the other two belonged, respectively, to Darwin and his servant Syms Covington. This information is confirmed from two sources.

First, there is a manuscript in Cambridge University Library that records, in Darwin's own spidery writing, that he returned from the

Galápagos with two juvenile tortoises. In the inventory of specimens, Darwin records:

Covington's little tortoise (Charles Island [Santa María])
Mine from James [San Salvador]

This is the only definite proof we have that Darwin and his servant actually collected their own juvenile tortoise specimens from the Galápagos. Darwin's scribbled note does not reveal whether his young charges are alive or not, but the fact that they are not listed as being among the *Beagle* reptile specimens donated to the British Museum in 1837 suggests that they probably were still living.

The evidence for the other two tortoises having belonged to FitzRoy comes from a different source. On 15 March 1837, the British Museum's Zoological Accessions book lists Captain FitzRoy as having donated seventeen animal specimens. Accession numbers two and three on the list are shown as being of the species *Testudo* from Española Island. These are almost certainly the two small tortoises whose measurements FitzRoy so carefully noted during the remainder of the *Beagle*'s voyage after leaving the Galápagos.

Unfortunately, it would appear that the captain may have lost interest in his tortoises after the end of the voyage, for the very fact that they were being donated to the British Museum's collections almost certainly meant that they had died in the meantime. The tortoises may have been cherished while on the ship but a combination of the British winter and FitzRoy's neglect may have sealed their fate once on shore. It is, of course, possible that FitzRoy's tortoises died long before the *Beagle* reached its home port. In his journal he does note that 'a very small [tortoise] lived upwards of two months on board without either eating or drinking', which perhaps means that after those two months it died. Either way, it seems that two of the four *Beagle* tortoises examined by Darwin and Gray were dead by March 1837.

That these four tortoises came from three Galápagos islands would seem to tally with Darwin's description of having examined tortoises 'that I brought from three islands'. Therefore, if Harriet is a remnant of Darwin's *Beagle* voyage, she was most likely one of the two tortoises collected by Darwin and Covington from San Salvador and Santa María islands respectively.

So far, so good; Harriet's story can be made to fit quite well. There were definitely some living tortoises brought back by the *Beagle* and they were young enough to tie in with Harriet's great age. However, there is still a large gap in time between 1837 and Harriet's earliest record of being in Brisbane during the 1870s. So what happened to the two tortoises after Darwin examined them in 1837?

～

Ed Loveday's 1994 letter only outlines a connection between John Wickham, the *Beagle* and the Botanical Gardens tortoises. He does not suggest that any of these tortoises were actually gathered by Darwin himself and then transported to Australia. However, since then the folklore surrounding Harriet has greatly increased and it is now usual for her to be described as a tortoise that actually once belonged to Darwin but which was later given to Wickham to transport to Australia. This is certainly a hypothesis that Scott Thomson has put forward and which has since been picked up by the international media. But how likely is this? Could Harriet really once have been Darwin's pet?

When trying to trace the fate of the two live tortoises that Darwin had in 1837, we encounter a paradox. Despite the hundreds of pages of notes and long lists of specimens that he made during the *Beagle* voyage, a comprehensive search shows that Darwin does not once mention that he and his servant took and looked after two live tortoises from the Galápagos. In fact the only reference to them at all is the previously mentioned note, after his arrival in England.

If John Wickham did take Harriet to Australia, he would have done so between 1841 and 1843, when he retired from the British Navy to become the police magistrate in Brisbane. This means that between 1837 and 1843, Harriet must have been living somewhere in England. Is it really credible that Darwin could have kept alive a tortoise for that length of time?

If one examines his life after his *Beagle* adventure, there seems to be little room in it for a tortoise. During the first six months he spent his time moving from place to place, alternately staying in Cambridge, London and Shrewsbury. When, in March 1837, he did settle in London it was as a guest of his brother Erasmus in Great

Marlborough Street. Erasmus Darwin was a socialite who mixed with an upwardly mobile group of philosophers and artists, and indulged in both drink and opium eating.

Erasmus' physical and intellectual extravagances would have been alien to the young Charles Darwin, fresh from a five-year voyage, and one wonders what he must have felt on entering the bohemian sector of London society. Not that it was all bad. It is quite possible that it was through Erasmus' friends that Darwin first became acquainted with the Reverend Malthus' theory of 'survival of the fittest', which in turn inspired his great concept of natural selection.

Darwin remained in Great Marlborough Street for nearly two years before he married his cousin Emma Wedgwood in January 1839. After this he took up residence in Gower Street, London. By this stage Harriet would have been four or five years old and would have been a reasonable size.

Given his habit of travelling frequently between various towns and cities, the likelihood of his being able to keep a pet tortoise seems remote, even if it did simply remain at his London house.

If Darwin had been living on his own, I would instantly discount the idea of his having looked after any of the *Beagle* tortoises for anything more than a few days after his arrival in England. However, he was not living on his own at this time.

In the days after his return from sea, Darwin's father (who had initially disapproved of his son becoming a ship's naturalist) was so pleased with Charles' success that he provided him with an annual allowance. One of Charles' first actions was to re-employ Syms Covington, the young man who had acted as his servant during the *Beagle* voyage.

Covington had remained aboard the *Beagle* after her arrival in England and, for several weeks after docking, was in charge of Darwin's specimens – including, one suspects, the two tortoises. When Covington left the ship to work for Darwin it is therefore quite probable that the tortoises went with him. This would certainly explain how they could still be alive and available to Darwin in the early part of 1837, when he took them to the British Museum to be studied by John Gray.

So, it was probably Syms Covington, rather than Darwin himself, who was in charge of the tortoises. One can better imagine Covington, the servant, having the time and inclination to tend to their needs than the young and fancy-free Darwin, preoccupied with theories of evolution.

This, however, does not entirely solve the problem, for Covington left Darwin's employment around May 1839, a few months after Charles' marriage to Emma. He emigrated to Sydney, Australia, where a glowing reference from his former master landed him a steady desk job. Given that Covington worked his passage on the long sea voyage to Sydney, it seems doubtful that he took with him two ten-year-old Galápagos tortoises (each of which would then have weighed around fifty kilograms).

It is at this point that the plausibility of Harriet's being Darwin's tortoise starts to waver. One has now to think of Charles and Emma Darwin, living in a tall London brick house with a small garden, a cook, at least two servants and, only a matter of months after the marriage, a child on the way. Would there really have been the space to look after two enormous tortoises? One has to be doubtful.

Then there is the question of Darwin's health. From the moment of his discovery of the theory of natural selection, he began to suffer from heart palpitations, breathing difficulties and other signs of sickness. These symptoms were to get steadily worse with time, sometimes forcing him to spend days resting in bed. Whether these bouts of illness were related to the stress of his research is unknown but their effect on the Darwins' family life was profound. Emma was often reduced to the role of nurse, as her husband was racked for days on end with vomiting fits, headaches and lassitude.

By the winter of 1841, Darwin was so ill that he could only work for a few hours each day. Again, faced with such problems, it is hard to imagine that the household could have coped with two ten-year-old tortoises as well as the two babies that Emma had by then produced. It is also hard to imagine Emma Darwin taking to the idea of having two giant tortoises in her home. She was very much in control of her household and, if the accounts are to be believed, ran a tight ship. Would such a Victorian matriarchal figure have put up with the giant tortoises? Again one must be doubtful.

The Darwins packed up and left London in December 1842 after what had been a particularly bad year for them. Aside from Charles' continued poor health, they had also witnessed a fully-fledged riot in the street below them and then, in October, lost their third child within a month of her birth. Disenchanted with London, the family retired to the Kent village of Downe where Charles' father had purchased a large house for them. There is no indication whatsoever that any tortoises moved with them, which suggests that, if they had been in his possession, by the end of 1842 Darwin had got rid of his two Galápagos pets. Quite where they had gone is another matter.

Trying to find out what happened to them between 1837 and 1842 has proved to be very frustrating. In this period none of the obvious London institutions, such as the British Museum, the Zoological Society or Kew Gardens, has any record of having received two Galápagos tortoises, alive or dead. Nor have any other major institutions, such as the museums at Oxford and Cambridge University. There are also no specimens in their collections that could match those collected by Darwin and Covington.

Trying to find evidence that Darwin held on to the two surviving *Beagle* tortoises has thus far proved fruitless. There is quite simply an absence of any evidence as to what Darwin did do with the two tortoises after his 1837 meeting with John Gray. Given this, I decided to tackle the problem from another angle and look into the claim that it was John Clements Wickham who had brought the tortoises to Australia from England.

John Wickham was first lieutenant on the *Beagle* when Darwin joined the ship and remained so for the duration of the voyage. There is little evidence that the two men became particularly close. On 5 July 1837, less than a year after Darwin's return to England, Wickham set off on the *Beagle*, again to survey the Australian coastline. The voyage was another prolonged venture, scheduled to last several years although Wickham was a crew member only until March 1841, when he was put on half-pay after apparently having suffered crippling bouts of dysentery for many months beforehand.

It is conceivable that Wickham could have taken Darwin's tortoises with him to Australia on this first voyage but, if so, there is no mention of the tortoises in the ship's records. Neither could he have left them in Brisbane, as the colony (then called Moreton Bay) was not officially opened for settlement until 1842. Moreton Bay in 1841 was a rough-and-ready frontier town of only a few dozen souls, so it seems more likely that if Wickham left the tortoises anywhere, it would have been in Sydney, Fremantle or one of the more established colonies. Again, the idea of his lugging two tortoises halfway around the world for several years before deciding to leave them in a backwater Australian town seems unlikely in the extreme. Therefore, if Harriet's provenance is to be believed, Wickham must have returned to England after 1841 and taken possession of the tortoises from Darwin (or someone else?) before returning to Australia. If he did not, Harriet cannot have been one of the *Beagle* tortoises.

Interestingly, Wickham's reason for leaving the Royal Navy might not be as clear cut as it looks on paper. During the *Beagle*'s two visits to Sydney, he had stayed with the family of Hannibal Macarthur during which time he fell head-over-heels in love with his daughter Anna. Now aged forty, Wickham wanted to settle down with Anna in Australia but first he needed a way out of the navy. The influential Hannibal pulled some strings so that his prospective son-in-law was invalided out. Wickham was still in Australia when he left the navy but when was he again in England?

He was certainly in residence in Sydney during March 1841 as his name appears on the census for that year. He was also probably in Sydney in April 1842, the date of a letter written by Lord Stanley, recommending him for the job of police magistrate in Moreton Bay and he was definitely there in October 1842 when he married Anna Macarthur. He took up his new position in late January 1843 after travelling there from Sydney aboard the *Shamrock*.

If Wickham returned to England from Australia after leaving the navy in March 1841 he would have to have made the return journey in a minimum of thirteen months (to be back by April 1842) or a maximum of twenty months, because he was clearly resident in Australia in October 1842. A direct passage to England would take at least three months (often more), which, after allowing for the process of organising the passage in the first place, would have left him little time for the round trip between March 1841 and either April or October 1842.

If Wickham returned to England in that period, it was an extremely brief visit. However, it is more likely that he never made the trip at all and remained in Australia. I have searched for evidence that Wickham was in England in 1841-2 but have found none. Whilst this is not proof in itself, it is nevertheless my belief that John Clements Wickham remained in Australia after he left the navy in March 1841. If so, the tortoises he allegedly donated to the Brisbane Botanical Gardens could not have travelled from England with him.

Aside from the fact that he probably did not visit England at that time, there is also no evidence that Darwin and Wickham ever communicated in the years immediately after Charles left the *Beagle* in 1836. Indeed, it was not until November 1862 that the two men met again, after Wickham had retired to Europe from Australia. This meeting was organised not by Darwin but by Bartholomew Sullivan, another of the original *Beagle* officers, and was part of a small 'reunion' of some of the old shipmates. In the interval, not a single letter or other communication seems to have passed between Darwin and Wickham. Many hundreds of people have raked over every detail of Darwin's life, but there is not one reference to the two men having kept in touch.

The idea that Wickham assumed ownership of Darwin's two tortoises without there being any correspondence at all, either to arrange the handover or to thank him, is almost impossible. Like many Victorian gentlemen, Darwin arranged his life by letter and kept an archive of all correspondence sent to him. To have no record at all of a meeting with Wickham, let alone his taking possession of two tortoises, is unlikely in the extreme. When all this evidence is looked at collectively, the case for Harriet having once come to Australia direct from Darwin's possession via Wickham is at best thin and at worst non-existent.

Add to this the fact that giant tortoises do not cope well with the English climate and also that most of the giant tortoises imported for zoos died within two years of their arrival (the longest a specimen has been known to survive is fifteen years and that was in the early twentieth century) and the case for Harriet having been one of the *Beagle* tortoises is weakened still further.

Assuming that Wickham was not responsible for Harriet's transportation leaves us with two main issues. The first is finding out what did happen to the tortoises that Darwin brought back on the *Beagle*. The second is finding out exactly where Harriet did come from.

The Fate of the *Beagle* Tortoises

A S WE HAVE seen already, the paper trail associated with the four known *Beagle* tortoises runs out in the spring of 1837. At this time we can be reasonably certain that FitzRoy's two tortoises were already dead and had been donated to the British Museum. We can also be reasonably certain that the other two tortoises, which were collected by Darwin and Covington, were probably still alive as they do not appear to have been donated to any museum. If this is so, what happened to them?

I have spent many, many hours searching for any clue to the possible fate of the *Beagle* tortoises. I have been through the archives of all the great Victorian institutions that might have received them and have pestered curators in universities and museums across the world, all in vain. It is as though the tortoises simply vanished from history after being examined by John Gray in 1837.

Then I had a lucky break. During one of my regular trawls through Darwin's own material, I came across a brief reference to a letter he wrote in 1874 to the great tortoise researcher, Albert Günther.

Although not an evolutionist, Günther had been a follower of Darwin's work for some time prior to their first communication in 1860. Indeed, Günther had been present at the Linnaean Society meeting in 1858 where the joint announcement was first made by Darwin and Alfred Russell Wallace of the theory of evolution through natural selection, paving the way for Darwin to write *The Origin of Species*. Günther was also in the audience in 1860 when the firebrand Thomas Huxley landed the first blow for Darwinian science against the anti-evolutionary rhetoric of Bishop Samuel Wilberforce.

It was the *Beagle* specimens that drew the two men together when, in 1860, Darwin spotted a paper that Günther had written on foreign snakes. Darwin, who had just published *The Origin of Species* and was in the middle of a very public debate on the question of evolution, made Günther aware of the snakes that he had collected from the Galápagos, perhaps in the hope that Albert would look at them and find more evidence in favour of natural selection. It was, however, not until the late 1860s that their relationship developed. Darwin was then engaged in writing *The Descent of Man* and again called upon Günther. More correspondence followed and, despite Albert's being twenty-five years younger than Darwin and dead set against the idea of natural selection, a friendship was struck up between the two. When, in 1869, Günther's wife died of a fever, Darwin wrote to comfort his friend: 'You have my most entire sympathy. Your words so full of tenderness and resignation brought tears to my eyes. I have at present no comfort for you, and I can only wish you fortitude to bear the greatest loss which a man can be given to bear.'

Günther, as we have seen, was obsessed with giant tortoises and, in the 1870s, made an attempt to trace the whereabouts of all those taken from the Galápagos, Mascarenes and Seychelles islands. During this quest he must have become aware of the *Beagle* tortoises and so naturally decided to write to his now ageing naturalist friend, asking him directly what had happened to his reptilian charges.

Darwin's reply was illuminating and, when I finally managed to track it down, it came as quite a shock to me.

> Down, Beckenham, Kent
> April 12 [1874]

My Dear Günther

I find that I did not bring home any tortoises from the Galápagos, as several were brought home by the surgeon and FitzRoy. I have vague remembrance that specimens were given to the Military Institution in Whitehall (where there is a larger model of the Battle of Waterloo) and I daresay Dr. Gray knows whether this keeps any specimens.

I am sorry that you were not able to come any day to lunch with us. I should have come and seen you again at the Museum; but unexpected business occupied *all* my time during my last week in London.

Yours very sincerely,
Charles Darwin

On the face of it Darwin seems to be denying outright that he had ever brought back any tortoises from the Galápagos Islands. This not only destroys Harriet's case completely but also seems to contradict the 1837 evidence that Darwin himself provided in the aftermath of the *Beagle* voyage.

In fact, despite his flat denial, we know that he did collect at least one tortoise. We have the evidence cited earlier, in Darwin's own handwriting, which specifically states that he and Covington collected tortoises from Santa María and San Salvador islands. When he wrote this letter, nearly forty years after the *Beagle* voyage, Darwin's memory must have been playing tricks on him. This is quite revealing in itself.

If Darwin had genuinely forgotten about his juvenile tortoises, this implies that he could not have had much involvement with them both during and after the voyage. If, as the legend of Harriet reiterates, Darwin and/or Covington had retained their tortoise(s) for several years whilst back in England, it is hardly likely that Darwin would have forgotten them (two giant tortoises wandering about the garden for several years would surely have stuck in his mind). Being good friends with Günther, he had no cause to lie about their existence.

This is confirmation to me that, whatever happened to Darwin's *Beagle* tortoises, they did not live with him for any length of time (if at all), yet there is no question that he still had access to them in early 1837 when he and John Gray examined them. Therefore at some point the tortoises were given away.

Darwin recalled that some of the *Beagle* tortoises were presented to the 'Military Institution in Whitehall', which owned a large model of the Battle of Waterloo. As matters stand we know of only four *Beagle* tortoises, and as two of these (FitzRoy's) were donated to the British Museum it is possible that the ones in Whitehall were those collected by Covington and Darwin.

The Military Institution concerned was the Royal United Services Museum, which occupied a large Whitehall building until its closure in 1968, after which many of the exhibits were dispersed around the world. Some of the Museum's archive has survived and was deposited in the Army Museum in Chelsea, London. I have searched this archive in the hope of finding a reference to the tortoises but without success. If the two unaccounted *Beagle* tortoises did end up there, which I believe is entirely possible, I have as yet been unable to find any trace of them.

With Darwin's own testimony against her, the case for Harriet having been one of the *Beagle* tortoises looks poor but there was yet more negative evidence to come.

In 1992, Scott Thomson enlisted the help of Scott Davis, an American scientist specialising in the genetic analysis of animals. Thomson had the worthy idea of trying to discover whether Harriet's DNA could reveal anything that would clarify her history.

Davis performed a mitochondrial DNA test on Harriet. The results, when compared to the DNA of other Galápagos tortoises, allow him to say with certainty which species she belonged to and thus which Galápagos island she was likely to have been taken from. The results concurred with Thomson's original identification that she was of the *Geochelone nigra porteri* subspecies from Santa Cruz Island.

This confirmation of her identification presents yet more negative evidence for Harriet being a *Beagle* tortoise as the ship did not call at Santa Cruz Island during its time in the Galápagos. Also, we know from Darwin and FitzRoy's notes that the four *Beagle* tortoises were collected from Española, San Salvador and Santa María islands, which, on the face of it, would seem to rule out Harriet as being one of them.

I put the question of Harriet's geographical incompatibility to Scott Thomson. It turned out that he too had been puzzled by this and he told me that, in his opinion, it was likely that Harriet had been brought to Santa María Island (where Covington collected his tortoise) from Santa Cruz by the prisoners in Governor Lawson's penal colony. 'My hunch,' he says, 'is that the prison colony was collecting tortoises from other islands for food and that Darwin collected some of these.'

It is true that, by the time of Darwin's arrival, the native Santa María Island tortoise was on the verge of extinction. Governor Lawson was therefore regularly sending parties of men to San Salvador Island in order to collect tortoises, process them and return with their salted meat. However, the men needed to be gone for several days at a time, so the islands to which they travelled must have had their own sources of fresh water. In this regard Santa Cruz poses a problem.

Frank Sulloway, an expert on the Galápagos and the history of the *Beagle* expedition, explained to me, 'It does not seem likely that the Vice Governor [Lawson] was sending people to Santa Cruz for tortoises, as there was no water there and the terrain is more impenetrable than on the other islands. There was water on Santiago [San Salvador], and the Vice Governor, for this reason, had sent people to this island to dry and salt fish, and to kill tortoises and salt the meat, evidently using salt from the salt crater at James Bay.'

So once again we have a difference of opinion and no firm means of determining the truth either way. However, the DNA analysis revealed more than just what type of tortoise Harriet is. It also provided data on her exact age. The biologist Scott Davis, who has analysed Harriet's DNA, told me the following:

> Harriet's nuclear DNA genotype at several microsatellite markers revealed some unique alleles not found in any other *G. porteri* (or other Galápagos tortoise, for that matter). Both of these facts are consistent with Harriet being a very old survivor of a larger, healthier population that had lots more genetic variation than the current tiny remnant. The hunting of the population to near extinction would eliminate a large portion of its genetic variation, and thus an animal removed over 100 years ago would be expected to carry a few unique variants that didn't survive the holocaust in the wild.

What this means is that Harriet was collected at a time when the Santa Cruz tortoises were still abundant and consequently had a much greater variety of genes than those born later, after the massacres of the whaling fleets. This supports the idea that Harriet was born some time in the first half of the nineteenth century and thus places her in the right timeframe for Darwin's visit.

This is logical, although in reality the tortoises on Santa Cruz did not suffer greatly at the hands of the whalers or the settlers (Townsend lists only 279 being taken between 1831 and 1868, compared to 4,326 from San Cristóbal); in the 1906 and 1974 censuses they were not listed as being endangered. Despite this, it cannot be denied that, based on Fleay's historical records alone, Harriet must have been born before 1850 and so it is not inconceivable that she was around in the days of Darwin.

~

So, at the end of all this research, are we any closer to solving the mystery of both Harriet's history and the whereabouts of the *Beagle* tortoises? In my opinion we probably are.

I am as certain as I can be that Harriet was not one of the tortoises brought to England by the *Beagle*'s crew in 1836. Aside from the complete lack of historical evidence in both England and Australia, she is from an island that was not visited by the *Beagle* and which was unlikely to have been visited by the Santa María Island settlers either. But what about the connection with John Wickham?

All the evidence against Harriet having been on the *Beagle* does not rule out the possibility that John Wickham did donate the tortoises to the Brisbane Botanical Gardens, merely showing that he did not obtain them from Darwin or any other member of the *Beagle* crew. If we reread Ed Loveday's letter, we can see that the information he was given in the 1920s simply notes that the tortoises were donated by Wickham. It is actually the retired historian Loveday himself who provides the connection between Harriet and the *Beagle*. So, perhaps Wickham did donate the tortoises after having obtained them locally.

As Scott Thomson's research (and Rothschild's records) show, visiting whaling ships were taking Galápagos tortoises to Australia during the nineteenth century. In my view Harriet is most likely to have arrived there in this way although there is only circumstantial evidence to back this up.

As for the four *Beagle* tortoises, we know that FitzRoy's two (dead) ones were donated to the British Museum. In my opinion the two belonging to Darwin and Covington probably ended up in the Royal United Services Museum where they would have succumbed quite quickly to the British climate.

Breaking the connection between Harriet and Darwin has been a great disappointment to me and it seems a shame to have to disprove her endearing story, but in many respects it was a tale too good to be true. However, Harriet was not the only pet tortoise with an interesting past. There is also the strange case of Marion's tortoise.

Marion's Tortoise

HARRIET IS AN emigrant from the Galápagos island of Santa Cruz, which has, at the last estimate, a remaining population of 2,000–3,000 tortoises. As such she is by no means a rarity but this is not true of all tortoise emigrants.

In his 1877 survey of the world's giant tortoises, Albert Günther concluded that Aldabra was the 'only spot in the Indian Ocean where the [giant tortoise] still lingers'. His assumption, based on historical records, was that the giant tortoises that had once existed on the other islands had become extinct in the very late 1700s or early 1800s. In Günther's mind, the Aldabra giant tortoise was now alone in its struggle for survival, precipitating his campaign to save the species.

However, in December 1893, Günther happened to read a piece in *The Illustrated London News* about a pet giant tortoise then living in the military barracks at Port Louis in Mauritius. The article, obviously a gap-filler on a day with little news, was mostly concerned with the animal's gigantic size and longevity. As an afterthought the journalist did note that 'it is just possible that the huge specimen now at Port Louis came from Rodrigues. Otherwise it may have come from Aldabra.'

To Günther this news was electrifying. By his own estimate the native tortoises of Rodriguez Island, located 1,000 kilometres from Mauritius, had vanished from the wild nearly a century previously. Could it be that the Port Louis tortoise was the last representative of the once-abundant Rodriguez Island tortoise race?

At the bottom of the article was an acknowledgement to a Captain S. Pasfield-Oliver who had provided much of the information. Günther lost no time in contacting *The Illustrated London News* and, on obtaining Captain Pasfield-Oliver's address, wrote to him directly.

He replied immediately and from his letter Günther was able to reconstruct the extraordinary history of what Captain Pasfield-Oliver called 'Marion's tortoise'.

The history of this noble beast begins in the 1760s with the adventuring Frenchman, Marc-Joseph Marion du Fresne, who worked as a sea captain for the colonial Compagnie des Indes. At the time Mauritius was firmly in the hands of the French and formed an important part of their Indian trade route. Marion du Fresne spent several years plying the Indian Ocean, stopping en route at various French protectorates. In 1766, Marion du Fresne called in at Rodriguez Island on his way back from the west coast of India. While there, like many other ships, he picked up some of the giant tortoises and took them on board. His ship moved on to call in at Port Louis, Mauritius, where for whatever reason one tortoise was offloaded and given to the soldiers at the barracks.

In the coming decades the French and the British battled it out with one another for dominance of the Indian subcontinent and the trade routes involved. Mauritius was a key staging post and in 1810, after having secured India, the British turned their attention towards the island, which was being used by the French as a base to harass British shipping. The initial attempt at invasion failed but in December 1810 an overwhelming British fleet overran the French garrison. The unfortunate tortoise, being housed in the artillery barracks, was right in the midst of the action and, minutes before the French surrender, received a direct hit on the top of its shell. The animal was injured, but not fatally, and for the rest of its days bore a noticeable dent in its carapace.

When the time came for the French to leave, they were reluctant to do so without their tortoise but the British, who were unwilling to concede an inch to their enemy, became stubborn. The French pleaded but to no avail. Marion's tortoise was to stay put and, to make sure of the matter, it was listed on the capitulation document, along with the remaining French ordnance, as specifically belonging to the British Royal Artillery.

Marion's tortoise, as it was known, was well cared for and as the British settled in the animal became a mascot for the Royal Artillery in whose barracks it was now stationed. In fact, the tortoise became

part of an unusual ritual in which each evening the commanding officer would sit astride its shell and be slowly carried to the mess hall.

Günther devoured this information but it was after Captain
Pasfield-Oliver provided him with a couple of photographs of
Marion's tortoise that he became convinced he was dealing with the
last-known example of a Rodriguez tortoise. The high dome and
smooth shell looked completely different to the flattened and highly
ridged shape of the Aldabra tortoise.

Marion's tortoise sunbathing in Mauritius in 1892. This tortoise was once
at the centre of a dispute between the French and English navies.

In a return letter to Captain Pasfield-Oliver, the aged Günther
confided that he believed the 'tortoise to be the Rodrigues species,
not one of the indigenous Mauritian species'.

Günther intended the comments to be private, not wishing to risk
his reputation on the identification of a tortoise that he had only seen
in photographs. Unfortunately, Captain Pasfield-Oliver had no such
scruples and immediately wrote a long letter to *The Times*, revealing
Günther's thoughts on the tortoise, and even quoting from a letter
the naturalist had written him in confidence.

Günther immediately put pen to paper, writing back to *The Times*:

> If I had known that Captain Oliver intended to publish my opinion as to the origin of the large tortoise living in Mauritius, I should have asked him to wait a few weeks, until we received some further positive information from his correspondents on the island . . . I am afraid positive evidence will be obtained only when, after the death of the animal, the bones can be compared with those of the other Mascarene tortoises. But I trust that great care will be taken in prolonging the existence of one of the oldest territorial creatures and, probably, the last survivor of its race.

Despite his public caution, in private Günther was absolutely certain that he was dealing with the last living example of a Rodriguez tortoise. The final proof came a few months later when he received photographs of Marion's tortoise tipped on its side, displaying its underside. Its distinctive characteristics were sufficient for Günther to declare to his friends that the tortoise was the last of its kind and that the greatest care should be taken with it.

Unfortunately, that care did not seem to be forthcoming. Captain Pasfield-Oliver wrote to Günther, expressing concern for its health. 'I fear,' he said, 'since they have shut up the old tort in a pen that he will not survive long. His original freedom was conducive to his age.' This pessimism seemed to be borne out when a short time later the tortoise was caught in a fire and, as a result, was permanently scarred.

Nonetheless, Marion's tortoise was actually to outlive Günther and was well treated in its old age. One soldier later recalled how he and his comrades had built a concrete bathing pond for it after the swamp in which it used to wallow was drained in the early 1900s. Unfortunately, the sides were too high so, whilst the grateful animal managed to get into the pond, it could not get out again. It took heavy lifting equipment and all the ingenuity of the Royal Artillery corps to remove the tortoise. A ramp was subsequently added so that it could come and go as it pleased.

By 1918, Marion's tortoise was suffering the effects of old age and was completely blind. This may well have been the cause of its ultimate demise. In mid-1929 it went missing but the soldiers were not especially worried as it was assumed that the animal was hibernating somewhere. By January it had not reappeared and a search was initiated. Its corpse was found at the bottom of a well. The blind tortoise had apparently walked straight into it and had been unable to get out again.

When the news was announced, *The Times* was flooded with letters from those who had seen and known the tortoise during their tour of duty on Mauritius. At this point Günther had been dead for fourteen years but the remains of Marion's tortoise were fished from the well and posted to the Natural History Museum in London where they reside to this day. Despite the affectionate tributes, it was still a sad demise for the last representative of an ancient Mascarene tortoise race and it now left only the Aldabra giant tortoise and those species still surviving on the Galápagos. At the time of the death of Marion's tortoise, nobody held out much hope for either and it was fully expected that within a few decades the world would be devoid of wild giant tortoises altogether.

PART SIX

Recovery

In the Name of Science

IN THE YEAR 1901, Lord Walter Rothschild, that avid collector of giant tortoises, was once again using his wealth to sponsor another collecting expedition to the Galápagos Islands. When asked to organise the expedition, Frank Webster, who had set up Rothschild's 1897 venture, commented drily, 'These creatures are so nearly extinct that any remaining ones will be only stragglers, and will only be secured at a great expense of time, hardship and money.' Undeterred, Rothschild employed Rollo Beck to head a new collecting expedition to the Galápagos in search of those few remaining tortoises that might previously have escaped his attention.

Beck had originally been employed by Webster as an ornithologist on Rothschild's 1897 expedition even though at the time he was just twenty-two. Now, with several more voyages under his belt, Webster felt that Beck had the necessary experience to head an expedition of his own. In addition, there were few volunteers prepared to undertake such trips. The schooner *Mary Sachs*, with Beck and company on board, duly left San Francisco in early January 1901 in search of yet more giant tortoises for Rothschild's collection.

Frank Webster's observation that there were only a few straggling tortoises left in the Galápagos was not just an idle comment. The whaling ships had abandoned the Galápagos some decades previously and the islands remained only sparsely populated, so the giant tortoise population should have been starting to recover. However, the dawn of the twentieth century was to bring an entirely new threat to the wildlife of the Galápagos Islands. This time it came not from ignorant sailors or settlers but, of all people, from scientists.

Since the early 1870s, Albert Günther had been gradually raising the scientific community's awareness of the plight of the giant tortoise but it was actually Walter Rothschild's 1897 expedition that made the zoological community take notice of just how few giant tortoises remained in the wild. In all spheres of life rare objects obtain a value that inspires people to want to own them. The same is true in science, and the obsession that drove Walter to want to collect the Galápagos tortoises in the first place was also felt by many academic institutions. To many museums, the size of their animal collections was a matter of great pride. Rothschild's boast that he had all but emptied Pinzón Island of its native tortoises sparked a desire in many academics for their own tortoise specimens. In the wake of Rothschild's expedition came a scramble by various universities and zoological societies to obtain giant tortoises before they became extinct in the wild.

The first scientific raiding party took place a year later in 1898 and was organised by Stanford University in California. The expedition spent six months in the Galápagos, bagging over 1,200 reptile specimens in the process, including a small, but unspecified, number of live tortoises from Pinzón and Isabela islands.

Although ostensibly a collecting expedition, the Stanford party also brought to the Galápagos a level of scientific observation that had not been seen since the visit of Darwin some sixty-five years previously. As well as being collected and dragged aboard ships, the tortoises and other fauna were observed in the wild for long periods of time, and their behaviour, habitat and habits noted in intimate detail. In the resulting report the two chief scientific collectors, Robert Snodgrass and Edmund Heller, were able to write the most detailed description of tortoise appearance and behaviour yet.

Whereas Darwin had had only a few weeks (just a few days of which were spent ashore) to observe the tortoises, Snodgrass and Heller could now describe the lives of the tortoises over a six-month period. During this time they noted how the tortoises liked to migrate to and from the highland areas using well-marked trails down the mountainside; how the tortoises obtained food and water during wet and dry seasons; and also their mating and reproductive habits. For the first time they also imparted something of the tortoises' personality, describing their love of mud baths, their mating battles and their dogged determination, of which the following is an example: 'They are very determined travellers, and once started in a certain direction no obstacles can stop them.

Not infrequently they ascend very steep, rocky hills. Sometimes their shells are broken, and occasionally they are killed, by rolling down these inclines, but if uninjured after these falls they will make repeated efforts to re-ascend until crowned by success.'

A year after the departure of Snodgrass and Heller, another expedition, under the control of Captain Noyes, brought back twenty-three tortoises from Pinzón and Isabela islands. Then came Rollo Beck's expedition for Walter Rothschild.

Despite the traditional plunder of the live tortoises, Beck also turned his hand to describing the animals but, unlike Snodgrass and Heller, he became acutely aware of the slaughter taking place around him. It was one incident in particular that opened his eyes to the perilous situation of the Galápagos tortoises.

Beck and his colleagues were following some of the tortoise migration trails up the Sierra Negra volcano on the southern end of Isabela Island when they came across a horrific sight. Beck had reached a flat, boggy part of the mountain, which, during the wet season, would fill with water and act as a drinking pool for the tortoises. However, instead of finding several tortoises slaking their thirst or wallowing in the shallow water, Beck was confronted by dozens of dead animals – over 150 by his estimate. The unfortunate creatures had evidently met with a grisly death, having been hacked to pieces with machetes, the marks of which were clearly visible in their shells and bones. What was surprising, however, was that these tortoises had not been killed for food, as their fleshly remains had just been left to rot, nor for their bones or shells, which had also been discarded. Quite why they had met their deaths was a mystery.

A little further on, surrounding another drinking pool were the remains of a further hundred tortoises, similarly butchered. As Beck's party continued, other slaughtered tortoises came to light. Ten here, another fifteen there. The death toll was enormous and perplexing. Surveying the carnage, Beck noted: 'At the rate of destruction now in progress it will require but a few years to clear this entire mountain of tortoises.'

It was only when he moved down from the mountain into the small coastal settlement of Puerto Villamil below that he found the cause of the butchery. In the space of the year or so since his last visit,

Hunters on Isabela Island would butcher tortoises using machetes in order to extract their oil.

a new industry had sprung up on the Galápagos – the collection and sale of tortoise oil.

An enterprising group of men from Ecuador regularly journeyed to Isabela Island (the only place where the tortoises were still common) to hunt them mercilessly for the few kilograms of fat lodged beneath the shells. Beck sought out and spoke to some of the men involved in this trade and was able to provide a vivid description of their activities:

> The outfit of the oil-hunter is very simple, consisting merely of a can or pot in which to try out the oil, and three or four burros for carrying the five- or ten-gallon [twenty-two- or forty-five-litre] kegs in which it is transported to the settlement. After making a camp near a water-hole, and killing the tortoises there, the collector brings up a burro, throws a couple of sacks over the pack-saddle, and starts out to look for more tortoises, killing them wherever found. A few strokes of the machete separate the plastron from the body, and ten minutes' work will clear the fat from the sides. The fat is then thrown into the sack, and the outfit moves on. When the burro is well laden, man and beast travel back to camp, where the oil is tried out. Each large tortoise yields from one to three gallons [4.5 to 13.6 litres] of oil. The small ones are seldom killed, because they have but little fat. By daily visits to the few water-holes during the driest season, in the course of a month the hunters get practically all the tortoises that live on the upper part of the mountain. When we first stepped ashore at the settlement we saw a number of casks lying on the beach, and learned on inquiry that they contained 800 gallons [3,632 litres] of tortoise oil. In a large boat, under a nearby shed, were 400 gallons [1,816 litres] more. While we were there, the boat sailing between the island and Guayaquil left for the port with those casks and a cargo of hides. The value of the oil in Guayaquil was about $US 9.00 per 100 pounds [45 kilograms]. While the tortoises are so plentiful as we saw them, this price yields a fair profit to the hunters, but two more raids such as that [which I encountered] will clear that mountain of all the fair-sized tortoises upon it, and then the oil business is ended.

Beck left the Galápagos with only a few animals but, on Walter Rothschild's instructions, returned in 1902, procuring another twenty-seven living (and twenty-three dead) tortoises, all of which were shipped straight out to the zoological park in Tring.

Rollo Beck was mightily troubled by what he had seen happening in the Galápagos. In his summary of the voyage, he was explicit about his belief that the tortoises' days were numbered: 'It is only within the last few years that the home of these very large tortoises has been invaded by man, but the rapidity with which they are being killed, and the reason for their destruction leaves us but little hope that they will survive any longer than did the American bison after the hide hunters began their work of extermination.'

Albert Günther had been making the same point for years but he had never visited the Galápagos and so could not speak with authority about the scale of destruction. Beck, on the other hand, had more personal experience of the islands and their giant tortoises than any other scientist alive and he knew that he was watching the wholesale extinction of an animal that had lived on earth for tens of millennia.

Beck was based in San Francisco, a wealthy trading city and the starting point for almost all the scientific expeditions that had visited the Galápagos in the early twentieth century. He knew that simply gathering tortoises and shipping them to England on behalf of Walter Rothschild was going to do little to further scientific knowledge of these animals. They had to be studied in the wild and for an extended period of time. Beck believed that there should be nothing less than a full-scale scientific expedition to the Galápagos that could properly document their wildlife before much of it disappeared for ever.

On his return to San Francisco, Beck made approaches to the California Academy of Sciences, an academic institution that had been founded during the California gold rush and which had rapidly become a university renowned for its research. Unlike most institutions of the Victorian era, the Academy had become aware of the environmental holocaust that industrialisation was bringing to many parts of the world. In particular, its scientists were concerned at the rate that many endemic island species were becoming extinct. It was with this in mind that, in 1903, the Academy launched its first ever island assessment expedition to the Revillagigedo Islands off the Mexican coast. The species there had come under threat from settlers clearing woodland for agricultural purposes. The Academy wanted to document the islands' plants and animals before they became extinct.

The Revillagigedo Islands expedition was a great success and so when Beck raised the possibility of a similar but extended survey of the Galápagos, it was taken seriously. In fact, the Academy had already been planning a return trip to the Revillagigedo Islands that included the nearby Cocos Islands. The Academy decided that it would make sense to extend this voyage so that it took in the Galápagos at the same time.

Such an undertaking was neither cheap nor easy to organise. The inhospitable nature of the Galápagos was legendary and for an expedition to be effective a highly trained crew of scientists would need to be away from home for a year or more. A large schooner was purchased and re-christened the *Academy*; over the space of several months it was fitted out so that it could be used as a comprehensive scientific base. For expedition leader there was really only one man with the scientific expertise and a wealth of first-hand experience in the Galápagos. Rollo Beck was duly appointed to head what would be his fourth and most exhaustive visit to the islands.

On 8 June 1905, after almost two years' preparation work, the *Academy* set sail from San Francisco, initially heading south to the Mexican islands before altering course westwards for the Galápagos Islands. As far as the tortoises were concerned, the expedition had two main objectives: (1) to find out the number of discernible species and the range of their distribution; (2) to try to work out where individual animals had originated. Both these questions had previously been addressed by Günther and, many years before that, by Charles Darwin, but satisfactory answers had not been found.

In 1877, based on the specimens available, Günther had believed that there were six separate tortoise species on the Galápagos. In the intervening years, George Baur and Walter Rothschild had added another five species, but because all these animals had been studied after their removal from their respective islands and by different people, it seemed quite probable that mistakes had been made. By observing the animals in their native populations, Beck hoped to be able to sort the jumbled mass of Galápagos tortoise species into some kind of order.

The expedition of eight scientists (including a geologist, botanist, entomologist, two ornithologists and two herpetologists) arrived in the Galápagos on 25 September 1905 and began their work. As they visited each island, the scientists would spread out, scouring the landscape, making notes, taking photographs and collecting specimens. The work was long and arduous but Beck wanted it done properly. The days accumulated into weeks, which in turn stretched into months. Finally, a year and one day after their arrival, the crew felt able to leave the Galápagos, having exhaustively searched every part of every island for its wildlife. The most comprehensive study of the Galápagos had been undertaken and the *Academy* was filled to the brim with rocks, minerals, plants, fish, insects, reptiles, mammals and just about anything else that could be caught and taken on board. The expedition's triumphant return was marred by the news that in their absence San Francisco had experienced the massive 1906 earthquake, levelling large sections of the city, including some of the Academy's buildings. Fortunately, the rebuilding effort had already been under way for several months and some level of normality was returning.

Naturally the *Academy* had a full complement of tortoises on board – 256 to be exact – collected from every conceivable Galápagos island and waiting to be handed over for formal study to the Academy's chief reptile researcher, John van Denburgh. However, it is ironic, given the expedition's aim of recording the tortoises prior to their extinction, that because of a lack of space in San Francisco, all the tortoises were killed and their bodies mummified so that they could be stored in the Academy's museum.

This act of clinical scientific brutality quite possibly resulted in the death of the last known tortoise from Fernandina Island, which Beck had captured after days of following its trail through the bush. A scientific expedition that had been despatched because of concerns about the fragile state of the Galápagos wildlife could quite possibly have rendered one species of tortoise extinct.

After the mummification had been completed, John van Denburgh spent many hours examining the new tortoise specimens, comparing them with each other and with existing descriptions, drawings and photographs. In the end he decided that he could discern fifteen species of Galápagos tortoise, including four new ones (see the table below). This was a major step forward in the study of the Galápagos tortoises and at last provided a firm base for future studies.

The key, however, was the geographical distribution of these fifteen species, which turned out to conform perfectly to Darwin's belief about island speciation. The various species of Galápagos tortoise were indeed restricted to very specific geographical provinces. Eleven were restricted to eleven individual islands but on the larger Isabela Island there were five species, each of which was restricted to a specific volcano. Evidently the geographical barriers (such as ravines, cliffs and craters) between the volcanoes prevented the five Isabela species from mixing with one another, keeping them genetically pure. This was a snapshot of the action of evolution on an island animal. Had he been alive, Darwin would have been delighted to see such emphatic evidence in favour of his theory of natural selection.

As it was, van Denburgh had made the greatest study of the Galápagos tortoises to date, surpassing even Albert Günther's 1877 work, and his fifteen designated species would remain unchanged through to the current day.

Island	Species	Status in 1906
1 Pinta	*T. abingdoni*	Rare
2 San Salvador	*T. darwini*	Rare
3 Rábida	*T. wallacei*	Very rare
4 Pinzón	*T. ephippium*	Fairly abundant
5 Santa Cruz	*T. porteri*	Not rare
6 Santa Fe	*T. ?*	Extinct
7 San Cristóbal	*T. chatamensis*	Nearly extinct
8 Española	*T. hoodensis*	Very rare
9 Santa María	*T. elephantopus*	Extinct
10 Fernandina	*T. phantastica*	Very rare
11 Isabela, Sierra Negra Volcano	*T. güntheri*	Abundant
12 Isabela, Cerro Azul Volcano	*T. vicina*	Fairly numerous
13 Isabela, Darwin Volcano	*T. microphyes*	Fairly numerous
14 Isabela, Wolf Volcano	*T. becki*	Fairly numerous
15 Isabela, Alcedo Volcano	*T. vandenburghi*	Rare

TABLE 1: A summary of the Galápagos giant tortoise species identified by the California Academy of Sciences expedition of 1906, together with their geographical range and estimated abundance. (From van Denburgh, 1914)

In more recent years the status of three of these tortoise species has come into doubt. The Fernandina tortoise (*T. phantastica*) is known from the solitary specimen pursued by Beck for so long. Since then, no other signs have been found of tortoises having lived on Fernandina, which is strange as it is the most untouched of all the Galápagos islands, so any native tortoises should have done well there. It is now thought that the tortoise captured by Beck may have been a lone male dumped there by a passing ship.

The same is true of the Rábida tortoise (*T. wallacei*), also known from a single specimen, which could conceivably have escaped from a stone pen used by whalers to hold tortoises from other islands. On Santa Fe Island the problem is a lack of material as the supposed species there is known only from a handful of fossil bones, the tortoises having apparently become extinct some decades before any specimens could reach scientific institutions. Although there is some doubt about these three species, for the sake of this book it will be assumed that they are valid until there is firm evidence to the contrary.

Sorting out the taxonomy of the Galápagos tortoises was welcomed by all those in the reptile field as well as being a major step forward in itself, but the Academy's expedition had some dismal conclusions about the status of the tortoises in the wild.

Although Beck and his team had not carried out an exact census of the tortoises, they had made estimates of their abundance in relation both to previous accounts and to their idea of a viable breeding population. The end result did not look promising for the long-term survival of the Galápagos tortoise. Of the fifteen newly defined species, two were thought to be extinct (those on Santa María Island and Santa Fe) while another six were classed as very rare or rare. Another five species were thought to be reasonably common, with only one of the Isabela species being listed as abundant. Although this rough census offered better news than the estimates of Beck, Günther and others, it was still not good. The numbers of tortoises were diminishing so fast that even the one abundant species was predicted to be in danger of extinction within a few years. The threat came not just from the indigenous human population and the commercial exploitation of the oil-collectors. There was another, far more serious, that no amount of educating the settlers was going to solve.

Beck had first noticed this during his 1902 visit:

> After seeing on this mountain dozens of tortoises of good size, one wonders where the small ones are but after spending a few days a-foot and seeing the many wild dogs in that region (descendants of those left years ago by sailing vessels) we can only wonder that so many of the large ones remain. From the time that the egg is laid until the tortoise is a foot long, the wild dogs are a constant menace, and it is doubtful if more than one out of 10,000 escapes. We certainly saw none, and the natives told us that the dogs ate them as fast as they were hatched.

On the 1905 voyage this menace from feral animals was much more obvious. Hatchlings and juvenile tortoises were found whose feet had been eaten by rats and dogs, while entire nests of tortoise eggs had been dug up and devoured almost as soon as they were laid. The Galápagos tortoise society was suffering from an inverse population pyramid: there were far more adults than there were hatchlings to replace them. Given the extreme age that the adults could reach, this meant that the giant tortoises could well have healthy populations for decades to come but, without any juveniles to supersede them, their numbers would slowly decline until, at some distant point in the future, they faded out altogether. The fact that the settlers and oil-collectors preferentially hunted the adults made matters worse. Joseph Slevin, the chief herpetologist aboard the *Academy*, believed that the Galápagos tortoise was racing towards extinction.

Like other scientists before them, the party from the California Academy of Sciences recorded all they saw and dutifully published it, complete with warnings about the future of the giant tortoise. Their contribution to the knowledge of the geology, geography and natural history of the Galápagos was the greatest yet, but unlike Günther none of the scientists was a campaigner by nature and so, after making the required ritual noises, the various members dispersed to follow their own paths (Beck later married and the couple toured the Polynesian islands, studying birds and taking photographs). The tortoises and other endangered Galápagos animals and plants were left to their own devices, still in want of a champion for their cause.

After this expedition, interest in the Galápagos as a whole was to die down completely for the next twenty years. It was only briefly revived in 1925 when Charles Townsend, the director of the New York Aquarium, undertook a survey of the effects of the whaling industry on the Galápagos tortoise population. Using nineteenth-century ships' logbooks, Townsend worked out that a minimum of 13,013 tortoises had been taken by the ships whose records he had studied. He was also able to rank the islands in terms of their loss of species. When compared with the estimated abundances given by the California Academy of Sciences' 1906 expedition, Townsend's figures tally rather well (with the exception of Isabela, where he could not distinguish between the five species).

Island	Tortoises taken	Abundance in 1906
San Cristóbal	4,798	Nearly extinct
Isabela	2,493	Fairly numerous
Santa María	1,775	Extinct
Española	1,698	Very rare
San Salvador	1,049	Rare
Pinta	455	Rare
Pinzón	356	Fairly abundant
Santa Cruz	366	Not rare
Santa Fe	23	Extinct
Total	13,013	

TABLE 2: Number of tortoises taken from the Galápagos Islands by whaling vessels between 1831 and 1868. (From Townsend 1925, and van Denburgh, 1914)

Given that there were dozens more whaling ships whose records Townsend had not studied, the true number of tortoises taken is likely to be nearer 100,000. It was no wonder that so few animals had made it into the twentieth century.

Like many other scientists before him, Townsend came to the conclusion that the tortoises faced extinction in the wild: 'The only remaining hope for the race is the establishment of survivors elsewhere.'

Like Albert Günther and Rollo Beck, Townsend was a man of action and five years later, on 1 March 1930, the last great scientific expedition for several decades departed from Miami, bound for the Galápagos Islands. Despite being in his seventies, Charles Townsend was on board and despite the relative brevity of the expedition in comparison to that of 1905–6, he still managed to gather nearly 180 live tortoises and to bring them back to the United States. This time they were not to be sacrificed in the name of scientific research or shunted to some millionaire's private collection. Townsend had much grander plans.

He had spent many years thinking about the problems of the giant tortoise and had come to the conclusion that the only means of saving them from total extinction was to establish breeding colonies in zoos around the world. The experience of Walter Rothschild's giant tortoises showed that they could not successfully breed in a cold northern climate and that they had to live in a tropical or subtropical environment in order to reproduce. Thus, rather than randomly distributing the tortoises about the world, he carefully selected zoos in San Diego, Hawaii, New Orleans, Bermuda, Panama and Sydney. The strategy worked and, whereas Rothschild believed himself lucky if any of his tortoises survived more than a couple of years once uprooted to Europe, many of those distributed by Townsend not only lived for several years but are still alive and well today. Some even managed to breed.

Even so, the conservation effort was a drop in the ocean and in many ways the continual taking of live tortoises 'for their own protection' was an admission by the scientific community that they were doomed in the wild. Unless some drastic action was taken on the Galápagos Islands themselves, the extinction of all the wild tortoise species was practically guaranteed. Right up until his death in 1944, Townsend kept campaigning about the plight of the tortoises, pleading with governments to intervene with either money or political pressure. The threat, he argued, was no longer the human population of the Galápagos but the feral animals. Exterminate them and the tortoises could be assured of some kind of future.

Unfortunately Townsend's campaign coincided with the build-up to the Second World War when governments had more on their minds than the plight of a few Pacific tortoises. His expedition of 1930 was to be the last serious interest taken in the Galápagos Islands for three decades. The tortoises, iguanas, birds, insects and other unique wildlife would have to struggle on against the odds until the world could once again wake up to the ecological value of these remote pinpricks of land.

Aldabra's Legacy

A T THE TURN of the twentieth century the Galápagos Island tortoises appeared to be on the steep and slippery slope to environmental catastrophe. However, they were lagging about a century behind the giant tortoise populations in the Indian Ocean, most of which had disappeared from the wild before 1800.

In fact, as far as scientists were aware, only Aldabra Island still supported any native Indian Ocean giant tortoises. In the 1870s, Albert Günther believed that the only way to save these was to transfer any remaining animals to a safe haven. In 1895 he achieved this when forty-two were taken from captivity in Mauritius and placed on Curieuse Island in the Seychelles. This was to be his last major intervention in the affairs of the Aldabra tortoise, after which he could only sit back and trust in its success.

As Curieuse Island was uninhabited, after their release the tortoises were, as promised, left to their own devices. Nobody is really sure what happened to them but they were all gone by 1970. A new population of ninety-five Aldabra tortoises was taken to Curieuse Island and released in 1978, but they did not fare well. By 1997 their number had only increased to 102 and many members of the original colony were missing. Under normal breeding conditions there should have been about 400 of them. Scientists blamed the shortfall on poachers and the low breeding rate on the island's population of rats and feral cats, as well as the native robber land crab, which would dig up and eat the eggs. If Günther's colony did not disappear because it was transferred elsewhere, it is, sadly, possible that the poachers, rats, cats and crabs caused their demise too. If the giant tortoises could not survive on Curieuse Island, what hope did they have on Aldabra?

The fear of extinction that drove Günther to set up the Curieuse tortoise reserve appeared to be justified when, in 1895, the Seychelles administrators undertook a comprehensive survey of Aldabra.

Amid a highly detailed breakdown of all Aldabra's wildlife (from its sharks to its flowers), there is barely a mention of the tortoises. Only on Picard Island, a small member of the Aldabra group where Mr Spurs had placed some tortoises a few years earlier, was there any sign of them. 'Traces of land tortoises' was the doleful comment, while other observations about 'very numerous' rats and a population of up to 400 goats seemed to confirm that the tortoises were fighting a losing battle to invading non-native species.

In the same year the naturalist William Louis Abbott spent three months living on Aldabra. In all that time he reported that he had 'met with but a few' tortoises and opined that there were now probably more Aldabra tortoises kept as pets elsewhere than there were on the island itself. Matters deteriorated further when, between 1895 and 1910, a succession of visitors to Aldabra reported seeing no tortoises at all.

Sir Walter Davidson, governor of the Seychelles in the opening decade of the twentieth century, gloomily wrote: 'No plan will effectively prevent the final extinction of these curious survivals in a wild state in their natural habitats.' John Stanley Gardiner, yet another British diplomat with a passion for wildlife, actually drove a route across one of the Aldabra islets in search of tortoises but found only two. He later commented: 'It would be possible to live for years on Aldabra and never see a specimen.'

Could it be that Günther's worst fears had come true? Was the Aldabra tortoise now extinct in the wild?

For a short while this was thought to be the case but not everyone echoed the pessimistic reports of Governor Davidson and others. The first signs of a change in fortune came in 1906 when naturalist Paul Dupont made a brief visit to Aldabra and noted that, contrary to what he had been told to expect, tortoises were 'still to be found in great numbers' in certain parts of the atoll. J. Fryer was more specific, citing that, while most of Aldabra was completely devoid of tortoises, they were still plentiful in two areas. One of these was Picard Island, where the tortoises released by Mr Spurs some twenty

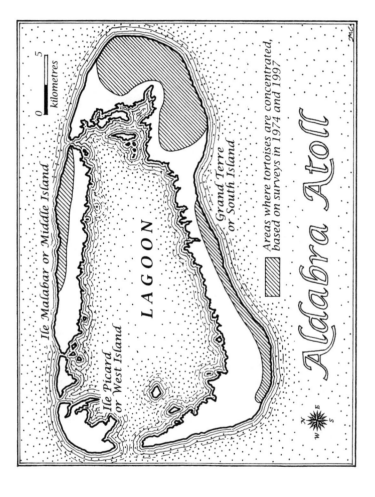

A map of Aldabra atoll showing the areas where the tortoises are concentrated.

years previously had obviously adapted to the island and had bred prolifically.

This improvement in the tortoises' chances on Aldabra may have begun in 1900 when the atoll was abandoned by the woodcutters. Because the trees were allowed to grow back, two of the tortoises' most precious resources, shade and leaves for food, were assured. That other precious resource, water, also became more plentiful without dozens of thirsty men competing for it. Shortly afterwards the Seychelles gained their independence from Mauritius and Aldabra Island was leased to Walter, Lord Rothschild, who forbade any commercial activity at all. It was during his time as landlord that the tortoises were able to make a small recovery in their numbers.

This is not to say that all commercial activity on Aldabra ceased for ever. Around 1908, Lord Rothschild's blackmailer engineered such a financial crisis that he had to terminate his lease. The tenancy went to the Seychelles Guano Company Limited, which started to exploit the island's reserves but at a much lower level than the woodcutters. The company continued to hold the lease into the late 1950s, during which time tortoise numbers continued to multiply prodigiously. (It should be noted that the Seychelles Guano Company paid a price for this exploitation: their workers on Aldabra were severely affected by malaria on a number of occasions.)

By the 1920s there were reports of hundreds of 'healthy looking' tortoises on Aldabra but less than a decade later their cause took a step backwards as the Seychelles government began exporting animals from the island to Kenya. Records indicate that between 1927 and 1976, over 1,100 tortoises were exported from Aldabra, probably either for food or to zoos. Surprisingly, the trade had no effect at all on the tortoise population and, with their numbers still on the increase, it looked as though Günther's worst fears had not been realised. Typically, however, Aldabra was not safe just yet. The greatest threat of all was yet to come.

∾

During the 1950s and 1960s, the United States had growing concerns about Soviet plans to dominate the Indian Ocean region and South-East Asia with their warships and military bases. The Americans, too, had ambitions in the area but were hampered by the limited fuel range of their jet fighter aeroplanes. A solution was needed and in 1964 pressure was put on the British, who were the major colonial power in the region, to find a suitable Indian Ocean island for a new US military airbase. In return the British would receive a subsidy of $11 million for the Polaris nuclear missile system, which they were then in the process of purchasing from the American government.

By April 1965, the British had drawn up a shortlist of island candidates. The final choice was between Diego Garcia, Cocos-Keeling and Aldabra. The Americas indicated that their first choice would be Aldabra, chiefly because it was unpopulated and therefore had no citizens to either object to the military plans or later claim compensation. Aldabra may not have had an indigenous human population to protest but, and largely thanks to Günther's work, it did have some highly placed and very vocal defenders.

The Royal Society of London, which had twice supported Albert Günther in his campaign to save the Aldabra tortoise, started the ball rolling in preventing the building of the military base. In July 1966, the Royal Society's Southern Zone Research Committee met to discuss the damage that such a base would cause to the delicate environment on Aldabra. It was during this meeting that the committee learnt of a forthcoming British government expedition to Aldabra. This was ostensibly to look at the possibility of erecting a BBC radio transmitter but there would also be a sizeable military presence in the party, supposedly to reconnoitre possible sites for the airbase. The Royal Society requested that two scientists be permitted to join the expedition. Permission was given and Charles Wright, of the London Natural History Museum, and David Stoddart, of Cambridge University, were despatched at short notice to the Indian Ocean, where they spent two weeks surveying Aldabra.

In 1895, the last full survey of Aldabra had painted a picture of an island in the midst of an ecological crisis. This time, the news was much better. The absence of man had allowed nature to take over once more, and the two scientists were delighted to be able to report that Aldabra was in pristine condition. Best of all, the giant tortoises were present in their tens of thousands (as were nesting seabirds). 'The atoll,' they surmised, 'is unique in the Indian Ocean.'

It was only on the expedition's return to Great Britain that the scale of the military's plans for their refuelling base became apparent. In order to be effective, the government said, the airbase was going to need a 4.5-kilometre runway, a deep-water harbour, aircraft hangars and crew housing. For an atoll that was barely thirty kilometres long and eight kilometres wide, most of whose area was taken up by a central lagoon, this would have been a disaster.

The military admitted that in order to accommodate the airbase, practically all the available land space on the island would have to be built on or altered in some way. The runway would be placed exactly where the greatest concentration of tortoises was found and would involve infilling all the freshwater pools used by both the tortoises and the birds. In order to create a deep-water harbour, some of the channels into the lagoon would have to be dammed, destroying the mudflats frequented by wading birds. Finally, there was the issue of all the foreign wildlife, seeds and diseases that would be imported wholesale with the troops and their equipment. After all, the non-native domestic animals and vermin had actually been the final *coup de grâce* for the tortoises in the Galápagos Islands. However it was viewed, the building of a military base on Aldabra would be catastrophic to the native wildlife, especially the tortoises, which would, by the army's own admission, have to be relocated elsewhere in order to survive.

In a report to the scientific journal *Nature*, Stoddart and Wright concluded that Aldabra was of the highest possible scientific significance, being the only atoll in the Indian Ocean in a completely natural state. Although they suggested ways in which the impact of any development could be limited, the real message was that any development at all was going to cause disruption on a massive scale.

The Royal Society was not alone in its concern about Aldabra. In another part of Britain, the British Seychelles Expedition Group was equally troubled. The group had also visited the island in 1964 and had identical fears about the negative impact of the airbase. Rather than attack on two fronts, the Royal Society and the British Seychelles Expedition Group teamed up to campaign for Aldabra. Their first action was to organise a conference in January 1967 in order to discus these issues with the wider scientific community and decide what course of action should be taken. Attendees included representatives from some of the most esteemed academic bodies in the world, including the Smithsonian Institution, the National Academy of Sciences and the London Natural History Museum, as well as from a dozen other universities, including Oxford, Cambridge and London.

Unusually for such academic gatherings, the delegates all spoke with one voice and agreed on three main points. First, that to build a military airbase on Aldabra would be 'a biological disaster'. Second, that a policy of total preservation was the only way forward for Aldabra. Third, that a research station should be established on the atoll in order to better understand its unique environment. These conclusions were released to the press, while a copy of the conference's findings was sent to the UK government with a letter requesting that the plan be reconsidered.

In the aftermath of the conference, news of the planned airbase began slowly to enter the consciousness of newspaper editors on both sides of the Atlantic. In March 1967, the *New York Times* opened the proceedings with an editorial on the subject. Its conclusion condemned the Americans and British for even suggesting that Aldabra might make a suitable military base. Soon the British broadsheets were also devoting column space to the issue, making the UK government feel distinctly uncomfortable with the level of press attention.

On 14 March, questions were asked in the British parliament regarding the scientists' concerns about the base. The Under-Secretary of State for the Royal Air Force tried to reassure both scientists and press with a statement that, typically for a politician, did not confirm or deny anything.

We have assured the Royal Society that the scientific considerations will be taken fully into account in the process of reaching a decision. We have also assured the society that if we have to construct an airfield on Aldabra it will be built in a way likely to cause the least interference with the ecology of the island, and that we would wish to extract some small virtue out of necessity not only by working closely with the Royal Society in conservation matters, but also in providing such facilities as we can for ecological and other studies. I acknowledge the deep concern felt by the Royal Society, and I wish to emphasise my Department's awareness of the issues and its willingness to do something about it.

Despite these assurances, it looked very much as if the politicians had already made up their minds in favour of Aldabra and had given more thought to ways of appeasing the Royal Society (among others) than to viable alternatives to the island.

Since it was clear that government thinking was in favour of building the airbase, the scientific community began to mobilise itself in a way that would have had Albert Günther smiling from ear to ear.

In the UK, the Royal Society put into action a plan for large parties of scientists to visit Aldabra in three successive phases, the first being that summer (the dry season in Aldabra), followed by another expedition six months later (the wet season), with a follow-up one at a later date. These would involve as many as eleven scientists at a time with the objective of recording in detail the natural conditions on Aldabra before it was covered in concrete and tarmac.

On the American side of the Atlantic, the scientific community began lobbying the government in Washington in the hope of changing attitudes. Tortoise expert W. Bourne vocally argued that Aldabra had only been chosen because it was a cheaper option than having to build on the other nearby populated islands. He believed that if the United States had to pay for the majority of the building work, the US government should be lobbied just as hard as its UK counterpart. In reply, the US Ministry of Defense said that it was well aware of the scientific issues and had discussed them with the British. However, the Pentagon took the line that the final decision did not rest there but with the British. No amount of lobbying was going to make the slightest bit of difference. The buck had been passed squarely to the UK, so that was where the scientific community had to focus its energy.

The Aldabra affair began to gather pace during the summer of 1967, a season in British terms when decent news stories are difficult to come by and the media are forced to look hard for headlines. A spat between the government and a group of distinguished scientists was just what the newspapers needed. Before long *The Times* and its companions were filled with the views of academics, all of whom were seething at the possible future destruction of Aldabra. Among the eminent names to speak out against the airbase were Sir Julian Huxley (descendant of Thomas Henry Huxley, Darwin's most vocal supporter), Viscount Ridley and Professor Wynne-Edwards. By the autumn, when the UK government reconvened after its summer break, the British public were well and truly aware of Aldabra's place on the globe and what the military had planned for it.

By October the more radical members of parliament were involved, asking awkward questions of the government about the island. One in particular managed to make more of an impact than most. This was the Scottish MP Tam Dalyell, who was well known for his interest in matters scientific and also for his ability to campaign aggressively on issues he considered important. Aldabra was one such and before the government knew what was happening Mr Dalyell was asking questions left, right and centre.

On 25 October, he stood up in the House of Commons and stated the Royal Society's scientific concerns in full, as well as his own more politically orientated views on the matter. The Under-Secretary again replied that no decision had yet been taken and that it was thought that some of the plans could be altered to alleviate the scientists' concerns. The Royal Society commented that this 'did not materially affect the case for preservation' that it had put forward.

Mr Dalyell was not happy. During the next month he tabled another fifty questions on the matter until he was banned from doing so by the Speaker of the House. At this point others assumed his mantle, with another dozen MPs tabling over thirty further questions within only a couple of weeks. Meanwhile, the second chamber of UK government, the House of Lords, scheduled its own debate on the matter for the end of November.

The media interest was also at a peak. An hour-long television programme was devoted to the issue and all the newspapers firmly backed the Royal Society. Only one periodical, *The Economist*, came out in favour of the airbase, noting that the scientific case was 'not unanswerable'. The list of academic institutions opposed to the plan was also growing as biological, ornithological, entomological, herpetological and other natural science departments and institutions around the world expressed their alarm. The British government was facing a serious public revolt but would ministers listen to the scientists and risk falling out with the Americans, or build the base and risk the wrath of the academic community? Before a decision could be made, the hand of fate intervened.

During the Aldabra debate, the British economy had been performing badly, stumbling from one crisis to another. One of the major problems was that the UK pound sterling was vastly overvalued in comparison to other major currencies, especially the US dollar. As winter drew near, an economic crisis looked ever more likely. To try to stave this off, the British Prime Minister decided, on 18 November, to devalue the pound.

The memory of this event is still fresh in the minds of many who were alive at the time, as it represented a public admission of defeat on the part of the UK government. In conjunction with this came a series of sweeping economic cutbacks on programmes that were considered too expensive to undertake in the late sixties. On 22 November, when the headlines were still screaming about the devaluation, it was quietly announced that, due to the looming national crisis, the plans for Aldabra were to be shelved. A week later it was admitted that, had it not been for the nosedive in the economy, the airbase on Aldabra would have been built as planned.

For some weeks afterwards the UK Ministry of Defence was keen to stress that the Aldabra airbase was only suspended while awaiting an upturn in the economy but as the months passed it became apparent that the plan was dead. Finally, the British government leased the Indian Ocean island of Diego Garcia to the American military so that their base could be built there instead. To avoid any logistical problems, the indigenous population of Diego Garcia was forcibly repatriated to Mauritius. Some people committed suicide rather than leave and there is still an active campaign by the islanders to be allowed to return to their native homeland.

The British and Americans got their military base and have used it to fly B-52 bombers from there to Iraq and Afghanistan in recent years. Given the way that the military treated the humans on Diego Garcia, it would appear that the Aldabra tortoises had a lucky escape. For the third time in less than a century, Aldabra had been the focus of a campaign by the scientific community to protect its tortoises. Thanks to the activities of the Royal Society among others, the world's last remaining wild population of giant tortoises had been saved.

The military threat to Aldabra had opened the academic community's eyes to the need to permanently protect the tortoises so that a similar situation could not arise in the future. In the weeks following the UK's climbdown over the military base, the Royal Society enacted their planned long-term research programme on the island. By 1971 it had founded a permanent research station on Aldabra and had implemented a programme of studies of its unique ecology and, more importantly, the welfare of its animals, especially the tortoises and nesting birds. Eventually the Smithsonian Institution, the United Nations and the World Wildlife Fund for Nature all contributed to the running and funding of this station. The reputations of these globally renowned institutions ensured that nobody would dare ever again to look at Aldabra as a development prospect.

One of the first pieces of good news discovered by the research station was the increase in the number of tortoises on Aldabra. A detailed census undertaken in 1973 revealed that there were approximately 129,000 living on the island. It seemed as though, at last, there was one small part of the planet where the giant tortoise was thriving and not under threat.

The Seychelles archipelago, which includes Aldabra, gained independence from Britain in 1976. Shortly afterwards the Royal Society's research station was closed down. In its place the new Seychelles government declared the island to be a special reserve that was strictly off-limits to all tourists, locals or anyone without a life-or-death reason to visit Aldabra. A smaller, low-key programme of scientific monitoring was set up, the main function of which was to make sure that the island's ecosystem was not interfered with by humans. A century earlier, Günther had believed that the best course of action for Aldabra would be to leave it alone. Evidently the Seychelles government now concurred with this and took pains to keep humans off Aldabra at all costs.

Aldabra's global ecological importance was recognised when, in 1982, UNESCO declared it to be a World Heritage Site, an honour that would ensure that any further ideas of developing the atoll would come under the scrutiny of the United Nations. In the same year, the dispossessed Diego Garcian people were given $7 million in compensation for their forced deportation although an attempt to further sue the British government failed in October 2003.

A test of the non-interventionist policy towards Aldabra came in 1997 when the Seychellois government permitted a detailed scientific survey of the island, including a follow-up to the 1973 tortoise census. This was carried out over several months and revealed that in twenty-five years the tortoise population had dropped from around 129,000 to about 100,000 – a decrease of 23 per cent.

However, the scientists did not find this decrease too alarming. A greater understanding of the tortoises' biology and habits had revealed a probable link between the levels of rainfall on Aldabra and the animals' ability to breed successfully. The 1990s had been a decade of particularly low rainfall so it was expected that the population would drop in response. The decreases could no longer be blamed upon the goats, pigs and rats; an extensive culling programme had practically eradicated these animals. At least 1,800 goats were killed in the 1980s so that, in 1997, only thirty were left on the island.

As it stands, Aldabra is the world's last uninhabited refuge for the giant tortoises and, barring some environmental disaster, their future on the island looks assured, at least for the short term. However, the more long-term problems of reported global warming trends and a possible rise in the sea level could yet threaten them. Unlike the volcanic and mountainous Galápagos, Aldabra is an atoll whose highest point is only a few metres above sea level. Even a small rise would cause the edges of the atoll to begin to erode, placing a renewed threat on the plant and animal life there. Also, if the periodic droughts that have affected Aldabra for the last two decades turn out to be a permanent trend, this too will have a large bearing on the tortoises. Indeed, there is now a debate among scientists concerning at what point humanity should step in and start to 'manage' the tortoise populations, rather than just leaving them to their own devices as is currently the case.

For the time being, however, we must be thankful for the hard work of Albert Günther, whose persistent badgering of the Seychelles and Mauritius colonial administration gave the Aldabra tortoises enough breathing space to ensure that they could survive into the twentieth century, until another generation of scientists could recognise their importance.

The Charles Darwin Foundation

S HORTLY AFTER THE Townsend expedition of 1930 left the Galápa-gos Islands, the world economy nosedived into the famous Great Depression. This catastrophe was swiftly followed by the turbulent events that led to the Second World War. The Americans and Europeans, who had organised and funded all the Galápagos scientific ventures, now had more pressing problems on their minds, and any concerns about the islands were forgotten for the time being.

At the start of the Pacific war between America and Japan, the Galápagos played a strategic role and for the duration of the war an American airbase was situated on the tiny islet of Baltra off Santa Cruz. In early April 1944, President Roosevelt and his wife Eleanor paid a visit to the troops there and were shocked at the islands' desolate appearance. Eleanor later wrote:

> To a geologist I'm sure [the Galápagos] would furnish several years of absorbing work, but to men establishing gun positions and defences, building airfields and trying to find a level space for a recreation field, it must be one of the most discouraging places in the world. It is as though the earth had spewed forth rocks of every size and shape and, as one man said, 'You remove one rock, only to find two more underneath.'

The American Army departed soon after the war but apart from a mysterious lack of iguanas, which were allegedly used by the troops for target practice, the islands do not seem to have suffered greatly at their hands.

During the 1930s, 1940s and 1950s, more settlers began to arrive at the Galápagos, establishing working ports and small fishing villages on the larger islands. The Ecuadorian government also set up a penal colony on the southern end of Isabela. It is known that the inmates were gathering the tortoises both for food and their oil. However, the fate of the population as a whole is largely unknown during this time.

It was not until the mid-1950s that another suitably qualified scientist set foot on the Galápagos. This was Iraenus Eibl-Eibesfeldt, a well-known European biologist. He took one look at the damage done to the plants and animals of the archipelago and immediately raised the alarm, warning that without prompt action much of the wildlife of the Galápagos would be wiped out within a generation. It was a message that had been relayed many times before and which normally fell on deaf ears. This time, however, it was to get a better reception.

During the late 1950s an international group of scientists began to lobby the Ecuadorian government about the state of the Galápagos. In 1959, the centenary year of the publication of *The Origin of Species*, the lobbyists established the Charles Darwin Foundation for the Galápagos and, after intense discussions, were permitted by the Ecuadorian government to set up a research station on the islands. It was to be another five years before it was open for business. In the meantime the members of the Charles Darwin Foundation began to undertake an extensive census of all the Galápagos Islands.

In 1906, when the last census of giant tortoises had taken place, only two of the fifteen island races were recorded as being extinct (those on Santa María and Santa Fe) but by the 1960s the Ecuadorians believed that this total had risen to eight, with all the remaining races being in a perilous situation. The first scientists arriving in the Galápagos were expecting the worst and so were pleasantly surprised to find that, whilst the situation for the tortoises was grim, it was not as bad as they had been led to expect.

The pessimism of the Ecuadorians had not been borne out. Of the thirteen species seen alive in 1906, ten were apparently still extant. This, however, was the end of the good news, for behind the hopeful-looking survival figures lay a crisis situation.

All the tortoise populations surveyed (with the exception of those on Alcedo Volcano, Isabela) were discovered to be in very bad shape indeed. Not only were they few in number but in almost every case

Island	Status in 1906	Status in 1960s
1 Pinta	Rare	None
2 San Salvador	Rare	Uncommon
3 Rábida	Very rare	None
4 Pinzón	Fairly abundant	Very rare
5 Santa Cruz	Not rare	Uncommon
6 Santa Fe	Extinct	None
7 San Cristóbal	Nearly extinct	Rare
8 Española	Very rare	Rare
9 Santa María	Extinct	None
10 Fernandina	Very rare	None
11 Isabela, Sierra Negra Volcano	Abundant	Rare
12 Isabela, Cerro Azul Volcano	Fairly numerous	Rare
13 Isabela, Darwin Volcano	Fairly numerous	Rare
14 Isabela, Wolf Volcano	Fairly numerous	Rare
15 Isabela, Alcedo Volcano	Rare	Common

TABLE 3: A rough estimate of the Galápagos giant tortoise numbers in the 1960s based on records made by the Charles Darwin Foundation. (1906 figures from van Denburgh, 1914)

the tortoises' society consisted solely of adults, with the young being entirely absent. The reason was easy to find: the rats, goats and dogs imported into the islands over the previous 150 years were eating the eggs and hatchlings. If the tortoises were left to their own devices, their extinction was not just a probability – it was guaranteed.

Faced with this reality, the first action of the staff at the research station, which was located on Santa Cruz Island, was to build a special tortoise-breeding centre. Here the eggs would be allowed to hatch in peace and the hatchlings grow to a decent size before being released back into the wild. The plan was a good one but although many observers had commented on the general behaviour of the tortoises, there was very little detail concerning their breeding habits.

It was known that the tortoises would usually mate during the Galápagos summer, especially between January and March. The pregnant female tortoises would then return to their normal routine until the onset of cooler weather in July triggered the urge to lay their

Breeding pens built by the Charles Darwin Foundation have been used to bring the Galápagos tortoise back from near extinction.

eggs. To accomplish this, the tortoises would wander away from the protection of the mountains, down to the lowlands, in pursuit of softer ground in which to construct a nest. Like those of most land reptiles, the nest is a shallow pit carefully dug using the hind legs and into which would be deposited somewhere between eight and twenty eggs, each about six centimetres in diameter. If their breeding programme was to work, the research station staff needed to find these nests before the rats and dogs.

The first expedition discovered that the soil disturbed by the tortoises during egg-laying made the nests clearly visible. The researchers cautiously dug down to the eggs, which were then carefully gathered and returned to the station for incubation, a process that normally takes a hundred days or more. They felt sure that success was guaranteed. It therefore came as a shock when, after weeks of patient nurturing, none of the first group hatched. What had they done wrong?

Trial and error uncovered the problem. It turned out that, in the early stages of their development, tortoise eggs are very sensitive to movement and even the slightest nudge could disrupt the growing embryo within. By harvesting the eggs within a few weeks of their being laid, the research station staff had unwittingly been killing the embryos. The eggs needed to be allowed to develop in the wild for at least two months before being gathered but this brought with it the likelihood that the goats and pigs would get to the nests first.

The only solution was to try to protect the eggs in the field. Initially this was done by piling rocks over the nest site to form a protective pyramid but some particularly determined pigs were still able to barge their way through. In time the primitive stone surrounds were replaced with swine-proof fencing.

After a given period, the protective rocks or fence would be removed from around the nest and the eggs carefully transported to the research station's artificial incubation chambers. Like everything else about the giant tortoises' reproduction, the biology of the egg incubation process was unknown and so the design of the incubators was subject to much refinement in the first few years.

Early designs tried to mimic the natural nest conditions by placing the eggs in shallow underground chambers but a low success rate led to several improvements. After years of experimentation, the eggs were incubated in a laboratory cabinet whose temperature could be kept constant to within one degree centigrade. Such close heat control was needed because, in the early 1970s, it was discovered that the sex of a tortoise hatchling is determined by the temperature at which the egg incubates. Eggs kept at 28° centigrade would hatch as males while those kept at 29.5° centigrade would hatch as females. This knowledge allowed the research station to bias its tortoise hatchlings in favour of females, the theory being that more females in a population would allow tortoise numbers to recover quickly. When the incubation process was mastered, approximately seventy-five per cent of all eggs gathered were producing healthy baby tortoises. Once hatched, their tiny shells would be numbered, and the hatchlings placed in ratproof pens. They would be carefully nursed through the early months of their lives before being transferred to larger outdoor pens to grow into adulthood.

Although collecting eggs from the wild was proving successful, on certain islands the total number of adults was so low that the tortoises could not find one another and so were not mating, let alone laying eggs. Particularly badly affected were the islands of Española and Santa Cruz where the extinction of both breeds was imminent. To counteract this it was decided to gather all the living adults from these islands and take them to the research station, where they could be protected from poachers and any eggs they laid could be instantly found and protected.

Gathering the adult tortoises was a huge logistical challenge and in the end a helicopter was hired for the purpose, lifting the gigantic animals beneath in large cargo nets. On their arrival at the research station, the different races of tortoise were kept apart from one another to prevent them from interbreeding.

The research station's actions were not without their detractors. Many people believed that removing the adult tortoises and their eggs from the wild would virtually guarantee their extinction in the wild. After all, it was by no means certain that any tortoises would ever be released back on to their native islands, and as a consequence the research station's programme would condemn them to the status of zoo animals.

This pessimism was not borne out and within only six years of its establishment the research station was able to release a batch of twenty Pinzón tortoises back on their native island. All twenty animals had been hatched at the station and reared there for five years until they had grown too big for the hungry pigs and dogs. Regular checks were made on the released tortoises and it was found that they had adapted to the wild without any difficulty at all, continuing to grow at a tremendous rate.

Releasing tortoises back into the wild was all very well but while the adults might have natural protection from feral mammals, their eggs and young were still vulnerable. Was there really any point in releasing adult tortoises whose ability to breed was seriously at risk?

The logical answer was to remove the feral animals from the Galápagos landscape but the Charles Darwin Foundation did not have the means (or, indeed, the permission) to begin such a massive culling programme. Pressure was brought to bear on the Ecuadorian government and in 1971 the first party of Galápagos park wardens was dispatched to Santa Fe Island to begin shooting the goats and pigs. In the years that followed similar culls were enacted on all of the smaller Galápagos islands. In one year alone over 26,000 goats were removed from the tiny island of Pinta. It was no wonder the tortoises had been having such a hard time of it.

In 1972 a goat-culling party on Pinta stumbled across a lone tortoise hiding in the highland region of the island. This was most unexpected as the Pinta race of tortoises was believed to have become extinct some decades previously. The discovery of this isolated tortoise raised the number of known surviving species from ten to eleven. However, the individual concerned was a male and, unless a suitable mate could be found for him, in time the Pinta giant tortoise would join the four other Galápagos races to have become extinct since the arrival of man. The tortoise was duly removed to the research station where he was given the nickname 'Lonesome George', allegedly after the actor George Gobel, who used the name during a television programme. The staff hoped that another Pinta island tortoise, preferably a female, could be found languishing in a zoo somewhere so that the race could be dragged back from almost certain extinction. To date no such mate has been forthcoming and Lonesome George continues to live his isolated existence at the research station, unaware that he is the last known representative of his species on earth.

Having fathomed the incubation process, by the 1980s the research station was producing hatchling tortoises at a stupendous rate. Even the adult Española tortoises, which numbered only fourteen individuals in 1968, were breeding successfully, allowing seventy-five subadults to be released back into the wild in 1975 alone. After two centuries of decline, the number of Galápagos tortoises was again on the rise.

The importance of the work of the Charles Darwin Foundation (and in particular its staff at the research station) cannot be underestimated. Without them the tortoises on Española, Santa Cruz and Pinzón islands would almost certainly be extinct by now, whilst the majority of the other tortoise races would be hovering on the brink of survival. In only thirty years this remarkable scientific effort has allowed over 2,000 sub-adult tortoises to be released back into a natural environment which, after much effort, has been restored to something like it was in the days before the whaling boats arrived.

Furthermore, their policy of openness and high-profile publicity has ensured that the world now knows of the ecological importance of the Galápagos Islands. As a consequence the Ecuadorian government tightly controls human development and settlement on the islands and is taking steps to ensure that its most famous national park can never again be subject to the wild excesses of mankind. These moves have come just in the nick of time, not just for the tortoises but for many Galápagos animals. We can only mourn those Galápagos species that have been lost for ever and hope that the mistakes of the past are not repeated although, given humanity's track record, this is far from being certain.

Thanks to the Charles Darwin Foundation and, in the Indian Ocean, the efforts of the Royal Society and other academic institutions, we have entered the twenty-first century with at least some of the world's giant tortoise populations still surviving, an achievement that looked most unlikely a century earlier. However, even by the late 1990s it was still not known just how many tortoise species had been saved and how many had become extinct. Despite centuries of debate, scientists could not agree among themselves as to how many species of tortoise there are (or had been) in the world. For a good while the problem appeared insoluble but help was to come from a most unexpected direction.

PART SEVEN

In the Blood

Lumpers and Splitters

B Y THE 1970s admirable efforts were being made to secure the Galápagos and Aldabra tortoises but from a scientific point of view one of the most enduring questions associated with these animals still had to be answered. How many giant tortoise species had there been on earth before mankind started rendering them extinct?

The last great surveys of the world's giant tortoise species had been made in 1914 and 1915, by John van Denburgh and Walter Rothschild respectively. Van Denburgh concluded that there had at one time been fifteen Galápagos species, a figure that Walter Rothschild endorsed the following year. From that time until the late 1970s, the idea that there were fifteen Galápagos species received few detractors. The only real change was to remove the Galápagos tortoises from the *Testudo* genus, which had been applied to all land tortoises since the 1750s, and place them instead in the smaller and more specific *Geochelone* category (meaning 'earth tortoise'). This aligned them with several smaller species of *Geochelone* tortoise found in tropical South America, Africa and southern Asia, reflecting the Darwinian belief that the Galápagos animals were the descendants of tortoises that had been washed up on the islands by accident.

The idea that there were fifteen species of Galápagos tortoise may have been endorsed by most zoologists but the same was not true of the Indian Ocean population. The early extinction of these races had caused confusion, leading Rothschild to identify seven species for the Seychelles Islands (including Aldabra) and another eleven for the Mascarene Islands.

Eighteen tortoise species for so few islands was a wildly over-optimistic figure and it looked as if somewhere along the line mistakes had been made. However, as the greater number of these species were now extinct and specimens were hard to come by, Rothschild was at a loss, admitting that he could not sort this mass of differing scientific opinion into a simple and coherent classification scheme. The problem was left to others, but few were willing to tackle it.

It was not until the late 1970s that another attempt was made to make sense of the species status of the Indian Ocean tortoise, when Nicholas Arnold of the London Natural History Museum completely reorganised the tortoises' taxonomy. After examining the seven species names used by Rothschild and others, all of which were within the genus *Testudo*, Arnold performed a task that scientists refer to as 'lumping'.

In place of Rothschild's seven Aldabra–Seychelles species, Arnold created just one, named *Geochelone gigantea*. Like the Galápagos tortoises, the *Geochelone* component reflected Arnold's belief that the Aldabra tortoise was descended from a closely related species from the nearby mainland, in this case Madagascar. The species name *gigantea* was a restoration of Schweigger's original 1812 name for the giant tortoises as a whole. By 'lumping' together under one name all the previously defined Aldabra and Seychelles species, Arnold had vastly simplified the chaotic taxonomy that once existed in the region. He also reiterated a common belief that the native tortoises on both islands were very closely related to one another.

Turning his attention to the Mascarene Islands, Arnold took the eleven previously defined species and lumped them into five distinct species from the three islands, which were then placed into the *Geochelone* genus.

Three of these species were originally named by Günther in the 1870s: *Geochelone inepta* (from Mauritius), *Geochelone triserrata* (Mauritius) and *Geochelone vosmaeri* (Rodriguez). The other two, *Geochelone indica* (Réunion) and *Geochelone peltastes* (Rodriguez), were older names that had been revived by others over the years and were thought to be valid still. Arnold's work had pared down Rothschild's eighteen species to six, all under the name *Geochelone*, but not everyone was content with this act of taxonomic lumping. It was time for the splitters to have their say.

Shortly after Arnold's revision came another reworking of the classification of Indian Ocean tortoises by the French scientist Roger

Bour who, in 1982, took a close look at *Geochelone gigantea*, the name for the Aldabra–Seychelles tortoise that had been the end result of Arnold's lumping. Bour was by nature a 'splitter', that is, someone who prefers to create new species by splitting them away from pre-existing ones. He was therefore the precise opposite of a 'lumper' and was thus inevitably going to reach different conclusions from Arnold.

The Aldabra–Seychelles tortoises are a classic example of how a group of animals can lead to strong disagreements between lumpers and splitters. To the layman, they all look alike, but to scientists, whose duty is to measure and describe every feature of these animals, there can be subtle differences in shell size, shape and other aspects of their biology. The issue is whether the differences between individual animals are strong and consistent enough to warrant separating them into different species. That is where the art of lumping and splitting comes into play.

Arnold, the lumper, acknowledged that there were numerous differences in shell shape between the many tortoise specimens that he had studied from Aldabra and the Seychelles, but he believed that these differences were due to diet and not genetics. He therefore ascribed all the variations in shell shape to the same species, *Geochelone gigantea*. This is analogous to the situation with the domestic dog, which ranges in size from the Afghan hound down to the chihuahua but, despite these variations, all are biologically the same animal and thus all are described under the name *Canis familiaris*.

Bour, the splitter, also analysed the degree of variation between individual Aldabra–Seychelles tortoises but decided that the differences in their shell shape warranted splitting them into five species. The new species were a mixture of pre-existing names and new ones given by Bour. For further good measure, he changed the *Geochelone* genus to *Dipsochelys*. In Bour's opinion, Arnold's single *Geochelone gigantea* genus could now be split into: *Dipsochelys elephantina* (from Aldabra), *Dipsochelys daudinii* (main Seychelles islands), *Dipsochelys hololissa* (main Seychelles islands), *Dipsochelys arnoldi* (main Seychelles islands) and *Dipsochelys sumierei* (Farquhar Island). But Bour was not the only splitter in this area.

A few years later two more scientists, Justin Gerlach and Laura Canning, reorganised the whole list, removing one of Bour's species

One species or three? These shells may look different to one another
but they all come from the single species native to Aldabra.

(*Dipsochelys sumierei*) and changing his Aldabra species from *Dipsochelys elephantina* to *Dipsochelys dussumieri*.

The issue became even more complicated when, in 1995, the Nature Protection Trust of the Seychelles received reports of 'odd-shaped tortoises' living in captivity in zoos and private residences on the main islands. On examining these animals, Gerlach and Canning believed that the animals conformed to their two new species, *Dipsochelys hololissa* and *Dipsochelys arnoldi*, which they had previously assumed to be extinct. By now the issue was really confused. Not only was there a surfeit of species but two of them appeared to have come back from extinction.

By the late 1990s, the situation regarding the Aldabra and Seychelles tortoises was as uncertain as it had been eighty years before. Was there one species or were there four or five? Were there three living species or just one? It seemed to come down to a matter of personal preference.

For a while there seemed to be no way of resolving the differences between the lumpers and the splitters until a revolutionary discovery in science came to their aid.

~

In 1985, Sir Alec Jeffreys invented a technique that allowed an individual's DNA to be mapped out in such a way that it could be quickly and easily compared with the genetic codes of other people. This technique was nicknamed 'genetic fingerprinting' and its uses were immediately apparent. Genetic fingerprinting was quickly adopted by the legal profession for use in paternity lawsuits, where a child's DNA could be compared with that of a supposed parent and thus either prove or disprove a genetic link. It was also used in murder and rape cases, where genetic material from blood or semen specimens from the crime scene could be compared with those of the suspect. But its uses were not restricted to the human world and very soon genetic fingerprinting techniques were also being used by zoologists.

Animals can be genetically fingerprinted and the results compared in exactly the same way as humans, and it was quickly realised that this technique could be useful in determining not just immediate relationships between individual animals (such as mother and daughter) but also the relationship between much larger groups of animals. Scientists began using genetic fingerprinting to work out the percentage of DNA that two animal populations had in common with one another. In this way they could work out the evolutionary relationship between them. For example, humans and chimpanzees share about ninety-nine per cent of the same genes, while between humans and gorillas the figure is abut ninety-seven per cent. This means that in evolutionary terms we are more closely related to the chimp than to the gorilla. (It should be noted that this does not mean that we are directly descended from either, simply that we share a common ancestor.)

On a smaller scale, genetic fingerprinting can also be used to determine small evolutionary differences between populations of animals that look very similar to one another. Returning to the canine analogy used earlier, dog populations around the world vary greatly in terms of their height, coat colour, physique, etc., but genetic fingerprinting tells us that despite these outward differences they are genetically identical to one another and thus all belong in the same species, *Canis familiaris*. The wolf, which looks superficially like a domestic dog, is, however, genetically sufficiently different to warrant being placed in a different species, *Canis lupus*.

In the mid-1990s it became apparent that the use of genetic fingerprinting to resolve the differences between animal populations could also be applied to the problem that the giant tortoises were posing to taxonomists.

The first attempt at genetically fingerprinting the giant tortoises was made in 1997 by the previously mentioned splitters, Justin Gerlach and Laura Canning, on a number of tortoises from Aldabra and the Seychelles, including the captive individuals that they believed to be examples of the once-extinct *Dipsochelys hololissa* and *Dipsochelys arnoldi*. When the results came back, they appeared to confirm that the two scientists had been correct and that, from a genetic point of view, there were indeed three living tortoise species on Aldabra and the Seychelles.

Based on this study, the Nature Protection Trust of the Seychelles began a new captive breeding programme to encourage the two newly discovered tortoise species to multiply. It appeared as though the splitters had won the argument – there were indeed several species of giant tortoise shared by the Seychelles and Aldabra – but the battle was not over yet.

In the 1990s genetic fingerprinting was a young science and there was still a moderate error range when making comparisons between animals. As time progressed, the accuracy of the tests increased until, by the turn of the twenty-first century, the error range was minute in comparison to only a few years previously.

In 2002 another group of scientists, which included Justin Gerlach, repeated the original 1997 genetic fingerprinting experiment on fifty-five living giant tortoises, using the newer, more accurate tests. The results were completely different. Instead of there being a neat division into three species, the DNA of all the animals tested was practically indistinguishable one from another and was certainly close enough to conclude that there was only one species of living tortoise and that was the native one from Aldabra.

These results were confirmed when another group of scientists, which included Nicholas Arnold (the lumper) and Roger Bour (the splitter), released the results of a completely separate genetic study of the Aldabra–Seychelles tortoise. Their study went even further than the previous one by including DNA that had been extracted from old museum specimens collected decades previously. Among these specimens were the tortoises that Bour, Gerlach and Canning had split off into new species a few years earlier. Would this new analysis support the idea that there had once been several genetically distinct species on the Seychelles?

The answer was no. As the previous study had shown, the genetic variation between all the Aldabra and Seychelles tortoises tested proved to be negligible. The conclusion was that the differences in shell shape were due simply to variations in diet and other local conditions. Despite their different shell shapes, underneath the tortoises were genetically identical to one another and thus all belonged to the same species.

In a dramatic turnaround, the lumpers had come from behind to win the day. The several species identified by Bour, Gerlach and Canning were duly dispensed with, the agreement now being that there had only ever been one species of giant tortoise on both Aldabra and the Seychelles. Under the complexities of taxonomic convention, the many names given previously to the Aldabra–Seychelles tortoise were swept away. This single species is now known as *Aldabrachelys gigantea*, which means 'the giant tortoise of Aldabra'.

After nearly two centuries of uncertainty, the complicated issue of how many tortoise species existed in the northern Indian Ocean had been resolved by the wonders of genetic fingerprinting. Now it was the turn of the Mascarene tortoises.

∽

The discovery that the shape of a tortoise's shell might not necessarily be the best way to define its species had implications for the status of the world's two other populations of giant tortoise. In the Mascarene Islands there were believed to have been five species of giant tortoise, all of which are now extinct, whilst in the Galápagos it was thought that there were once fifteen species, of which four are extinct. However, this allocation of species names had been entirely based on the animals' physical characteristics, especially their shell shape. What if the scientists had been too enthusiastic and, like the splitters with the Aldabra–Seychelles tortoises, had created several species where there was in fact just one? There was only one way to find out and that was to repeat the genetic fingerprinting experiment on these tortoises too.

It was here that the Mascarene tortoises presented the greatest challenge, as every one of the five proposed species had been extinct for over a hundred years. All the scientists had to work with were a handful of preserved bones, sitting in the archives of various museums around the world. Fortunately DNA is resilient, and provided that the skin, bones or shell of an animal have not been subjected to any chemical or other extreme preservation process, there is normally enough genetic material remaining to permit an analysis.

In 2001, Jeremy Austin and Nicholas Arnold managed to extract DNA from several museum specimens, the youngest of which was 170 years old and the oldest over a thousand years old. Despite the age of the specimens, Austin and Arnold managed to get DNA samples from all five of the supposed Mascarene species. The results were quite unexpected.

Unlike the studies of the Aldabra–Seychelles tortoises, the genetics of the Mascarene tortoises confirmed that the scientists were correct in their identifications. There had indeed been five genetically distinct populations of tortoises living on the three islands. Furthermore, a quick comparison between the genes of the Mascarene tortoises and those of the species from Aldabra (*Aldabrachelys gigantea*) showed that there was a big difference between the two. To recognise this difference, the Mascarene tortoises were all grouped together under the new name of *Cylindraspis*.

The names and locations of the five Mascarene species were exactly as Nicholas Arnold had predicted back in 1979. Native to Mauritius were *Cylindraspis inepta* and *Cylindraspis triserrata*; in Rodriguez were *Cylindraspis vosmaeri* and *Cylindraspis peltastes*, while *Cylindraspis indica* was the sole species on Réunion. With all the Indian Ocean tortoises genetically fingerprinted and correctly speciated, that left only the fifteen species from the Galápagos Islands. How would these fare under the scrutiny of the geneticists?

By the time the geneticists came to study the Galápagos populations, its fifteen species had been revised so that they were grouped together under the single species *Geochelone nigra*. The fifteen former species had been designated a subspecies status (for example, what had been called *Geochelone porteri* became *Geochelone nigra porteri*). This change reflected the increasing belief among scientists that the Galápagos tortoises were much more closely related to one another than had previously been believed. Would genetics bear out this belief?

Because of the fame of the Galápagos tortoises, more than one group of scientists set out to study their genetics, with the results being published between 1999 and 2003. The aims and objectives of these studies varied slightly but the results broadly coincide. DNA samples were taken from all eleven of the living species and analysed. Due to a paucity of material, this has not as yet been done with the four extinct species – *Geochelone nigra wallacei* (Rábida Island), *G. n. elephantopus* (Santa María Island), *G. n. phantastica* (Fernandina Island) and the unnamed species from Santa Fe Island.

~

The results of the genetic analysis bore out John van Denburgh's 1914 division of species. The eleven tortoise populations analysed were found to be very closely related but there was enough genetic

variation to justify their subdivision into eleven subspecies. One
study concluded that the four subspecies on southern Isabela Island
(*G. n. microphyes, vandenburghi, güntheri* and *vicina*) were genetically
so close to one another that there was a case for them to be merged
into one subspecies. However, at the time of writing this has not
been done and the Galápagos are still considered to have fifteen subspecies as defined by van Denburgh in 1914. Needless to say, a comparison between the genetics of the Galápagos tortoises and those of
the Indian Ocean showed a great gulf between the two, proving that
one population was not descended from the other.

Thanks to all these genetic surveys, one of the most outstanding
questions associated with the giant tortoises has been answered. We
can now say with reasonable certainty that within historical times the
following tropical islands have held populations of the following
giant tortoise species:

ALDABRA AND THE SEYCHELLES
Aldabrachelys gigantea (living)

MASCARENE ISLANDS
Cylindraspis inepta (Mauritius, extinct)
Cylindraspis triserrata (Mauritius, extinct)
Cylindraspis vosmaeri (Rodriguez, extinct)
Cylindraspis peltastes (Rodriguez, extinct)
Cylindraspis indica (Réunion, extinct)

GALÁPAGOS ISLANDS
Geochelone nigra becki (Isabela Island, Wolf Volcano, living)
Geochelone nigra microphyes (Isabela Island, Darwin Volcano, living)
Geochelone nigra vandenburghi (Isabela Island, Alcedo Volcano, living)
Geochelone nigra güntheri (Isabela Island, Sierra Negra Volcano, living)
Geochelone nigra vicina (Isabela Island, Cerro Azul Volcano, living)
Geochelone nigra darwini (San Salvador Island, living)
Geochelone nigra abingdoni (Pinta Island, one surviving individual)
Geochelone nigra ephippium (Pinzón Island, living)
Geochelone nigra porteri (Santa Cruz Island, living)
Geochelone nigra chatamensis (San Cristóbal Island, living)
Geochelone nigra hoodensis (Española Island, living)
Geochelone nigra wallacei (Rábida Island, extinct)
Geochelone nigra elephantopus (Santa María Island, extinct)

Geochelone nigra phantastica (Fernandina Island, extinct)
Geochelone nigra? (Santa Fe Island, extinct?; known only from a few
   bones and yet to be officially named)

All this genetic work confirms what Darwin had first suspected all those years ago – namely that the giant tortoises of the Indian Ocean are different from those in the Galápagos. Furthermore, within the Galápagos themselves the different islands do indeed have their own endemic species of tortoise, just as Governor Lawson had stated. Only one further question remained. Where did the giant tortoises come from in the first place? Genetics would be able to help here too.

The Tortoises' Origins

THE ORIGIN OF the world's giant tortoises was not perhaps the hottest topic in nineteenth-century scientific circles but a few people were vexed by it. Those who did discuss the question generally fell into two groups. The first consisted of people like Albert Günther, who believed that the tortoises had not evolved on their remote islands but had somehow managed to get there as fully formed species and had remained *in situ*, unchanged. The second group generally consisted of pro-Darwinians, who preferred to believe that the tortoises had been carried by the sea to their remote islands and, once established there, had adapted themselves to the local environment, in the process producing new species.

By the mid-twentieth century it was generally agreed that Darwin had by and large solved the enigma of evolution and that animals like the giant tortoise had evolved into new species *after* they had been swept on to their remote islands. Nonetheless, quite where the tortoises' ancestors had originally emigrated from and exactly where and when they had first landed on the islands continued to be a mystery. The science of genetics was about to provide an answer.

As with determining the number of species, the key to tracing the evolutionary history of the giant tortoises came from measuring the genetic differences between their various species. As with the example of humans, chimpanzees and gorillas cited earlier, the more genetic material that two species have in common, the closer they are in evolutionary terms. By comparing the genetic gap between the various tortoise species (or, in the case of the Galápagos, subspecies), it is possible to see the order in which the species evolved and who their closest living relatives are. Also, because evolutionary changes take place at a relatively consistent pace through time, it is possible to look at the differences between two genetic populations and work out approximately when their common ancestor was alive.

This information was used to construct a rough family tree for the three main giant tortoise populations around the world and also to estimate when in geological history the various branches split off from one another. Before these family trees could be drawn up, the scientists also genetically fingerprinted a number of other living tortoises in East Africa, Madagascar and Central and South America. This was so that they could find the closest genetic match to the giant tortoises on the oceanic islands.

The first evolutionary family tree to be constructed was that of the fifteen Galápagos tortoises. Using genetics, the scientists determined that the closest living relative to the fifteen subspecies there was the Chaco tortoise (*Geochelone chilensis*), which resides in Peru and Chile. The Chaco tortoise is actually quite small with a shell length of only forty centimetres or so and thus quite different to its gigantic cousins on the Galápagos.

The genetic distance between the Galápagos and Chaco tortoises indicates that the two species probably shared a common ancestor around 6 to 12 million years ago. So, several million years ago, the Galápagos tortoises' original ancestor was swept out to sea from the coastal region of southern Peru or Chile. After being caught in a strong offshore current, this ancient tortoise would have floated for a week or so until, by chance, it was washed up on to one of the Galápagos Islands. But what happened then?

According to the genetic differences within the eleven Galápagos subspecies that were tested, the oldest division into subspecies took place around 2 million years ago. This tells us that the tortoises were already on the islands at this time, which fits in well with the known geological ages of the various Galápagos Islands, which range from 5 million to 500,000 years.

So, some time between about 5 and 2 million years ago the ancestor (or ancestors) of the Galápagos tortoises were washed up on a beach, having crossed 1,000 kilometres of open ocean from South America. This ancestor was probably washed up on one of the islands in the south-east of the archipelago, such as Española or San

Cristóbal, as these islands are geologically the oldest and also because genetic analyses show that the tortoises from these islands are closely related to one another while being more distantly related to the rest of the Galápagos tortoises. From the south-east Galápagos the tortoises gradually spread to other islands, probably by being washed off the shore of one island only to be carried by favourable currents to another.

From Española or San Cristóbal, the tortoises were carried to Pinta, Santa Cruz and San Salvador Islands and from there to Pinzón and Isabela islands. In fact, Isabela, which at 500,000 years is the youngest of all the Galápagos Islands, had been colonised twice by the tortoises. In the south, Isabela was invaded by tortoises swept in from Santa Cruz Island, which then diversified into the four subspecies associated with its four volcanoes. In the north of Isabela, another tortoise arrived from San Salvador Island and evolved to become the *becki* subspecies that is associated with Wolf Volcano. During this 2-million-year traffic between islands, the isolation of the tortoises' existence led to sufficient genetic differences for scientists to divide them into fifteen different subspecies.

Unfortunately, as we have no genetic information from the extinct tortoises from Rábida, Santa Fe, San Salvador and Fernandina islands, it is difficult to know exactly where these tortoises fit into the scheme of things but, based on the position of the islands, it seems likely that the Santa Fe and Santa María tortoises came from Española or San Cristóbal, while the tortoises on Fernandina probably came from Isabela.

Some 165 years after Darwin's visit to the Galápagos, the evolutionary history of the islands' giant tortoises had been clarified and the results conformed perfectly to his theory of natural selection. Island isolation did indeed play a crucial role in the evolutionary origin of the Galápagos giant tortoises. Other genetic analyses would confirm that the same evolutionary processes had also operated on the finches, creating distinctive species for each island.

~

The story of tortoise colonisation in the Indian Ocean was another matter altogether. The five extinct species of Mascarene tortoise,

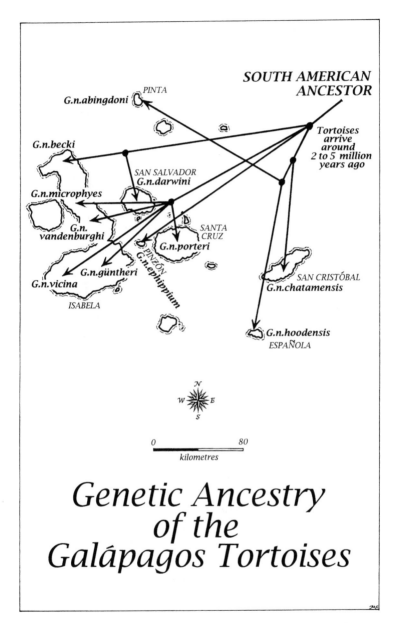

SOUTH AMERICAN
ANCESTOR

PINTA
G.n.abingdoni

G.n.becki

Tortoises
arrive
around
2 to 5 million
years ago

SAN SALVADOR
G.n.darwini

G.n.microphyes

**G.n.
vandenburghi**

*SANTA
CRUZ*
G.n.porteri

PINZÓN
G.m.ephippium

G.n.günther i

G.n.vicina

SAN CRISTÓBAL
G.n.chatamensis

ISABELA

G.n.hoodensis
ESPAÑOLA

N
W   E
S

0   80
kilometres

Genetic Ancestry
of the
Galápagos Tortoises

The evolutionary relationship between the living species of
Galápagos tortoise has been revealed through their DNA.

whose DNA had been extracted by Austin and Arnold, turned out to have had a relatively simple history. A genetic comparison between the Mascarene tortoises and the living tortoises on Madagascar has yet to be made, but there are fossilised giant tortoises on Madagascar that are morphologically similar to the Mascarene tortoises. Based on this, it is assumed that it was on Madagascar that a tortoise first stumbled and was swept out to sea. From this point genetics can pick up the thread.

Several million years ago one of these Madagascan giant tortoises (or perhaps several of them) was carried out to sea and rode an eastward current for 800 kilometres to the island of Mauritius. There it thrived, evolving into two new species, *Cylindraspis inepta* and *Cylindraspis triserrata*. The latter of these species settled down on Mauritius and never left. *Cylindraspis inepta*, however, seems to have been accident prone, and individuals from that species were twice swept off the coast of Mauritius, one drifting west to Réunion and the other east to Rodriguez.

The newly arrived *Cylindraspis inepta* tortoise on Réunion stayed there, eventually evolving into a new species, *Cylindraspis indica*. The Rodriguez *Cylindraspis inepta* tortoise, however, managed to evolve into two separate species, *Cylindraspis peltastes* and *Cylindraspis vosmaeri*. That was how the situation remained until the sixteenth century when all five species came into contact with humanity and were wiped out.

The history of the single giant tortoise species of Aldabra and the Seychelles, *Aldabrachelys gigantea*, was different again. Prior to the genetic surveys, opinions had differed not only on the number of species of giant tortoise in those islands, but also on their evolutionary history. Some said that the tortoises had originated on Madagascar, and had been swept to Aldabra and from there to the Seychelles. Others said that they had been swept from Madagascar to the Seychelles and from there back to Aldabra. Like the number of species, it was a riddle that would only be solved with the arrival of genetic analysis.

As was shown in the last chapter, the genetic survey of the tortoises of the Seychelles and Aldabra showed that they were all very closely related to one another and were thus grouped together under just one

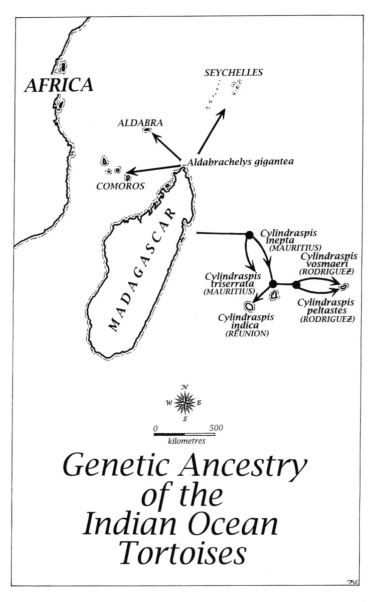

AFRICA

SEYCHELLES

ALDABRA

Aldabrachelys gigantea

COMOROS

MADAGASCAR

Cylindraspis inepta
(MAURITIUS)

Cylindraspis vosmaeri
(RODRIGUEZ)

Cylindraspis triserrata
(MAURITIUS)

Cylindraspis peltastes
(RODRIGUEZ)

Cylindraspis indica
(RÉUNION)

N
W E
S

0   500
kilometres

Genetic Ancestry of the Indian Ocean Tortoises

The dispersal route and evolutionary relationships between the living and extinct species of Indian Ocean giant tortoise, as revealed by genetic analysis.

species, *Aldabrachelys gigantea*. Therefore, unlike the Mascarene and Galápagos tortoises, there is little genetic variation between them, making their evolutionary history difficult to construct. Fortunately, other clues were available.

Like the Mascarene tortoises, the Aldabra tortoise species also has a strong physical similarity to fossil species found on Madagascar, so it is likely that this is where their ancestors originated. The ocean currents from the northerly part of Madagascar travel directly northwards towards Aldabra, the Seychelles and the Comoros islands. *Aldabrachelys gigantea* certainly lived on the first two island chains in this list, whilst fossilised bones probably belonging to this species have been found on the Comoros. This suggests that the original Aldabra tortoise may well have lived on the north of Madagascar and been swept out to the other islands from there. Trying to work out when these migrations happened is more problematic.

Because of the close genetic relationship between all the Aldabra and Seychelles tortoises tested, clearly they have not had much time to diversify and therefore their migration from the mainland must have been relatively recent. Due to the high sea level during the last ice age, Aldabra was underwater until about 125,000 years ago, suggesting that the tortoises' migration from Madagascar is more recent than that. Of all the giant tortoises in the world, it would appear that the species on Aldabra was the most recent to leave its mainland home and settle on the remote oceanic islands.

~

The revelations provided by the recent genetic studies of the world's giant tortoises have been profound, clearing up the last outstanding mysteries to surround these animals since scientists first took an interest in them over two centuries ago. Thanks to genetic science we now know how many giant tortoise species there once were on our planet and, conversely, how many mankind has managed to render extinct. We have also managed to understand something of their origins. Hopefully there will be further revelations to come as scientists improve in extracting DNA from fossil material, and as some of the many giant tortoises languishing in zoos worldwide are tested.

The giant tortoises have the ability to inspire, amuse and amaze, and there may yet be a few surprises hidden beneath the shells of these gentle but persecuted giants.

Epilogue

A T THE TIME of writing, Lonesome George is the sole surviving member of the *Geochelone nigra abingdoni* subspecies that once populated Pinta Island in vast numbers. Despite many efforts, which include a $10,000 reward, all attempts at finding him a suitable mate have thus far failed. Even attempts at getting him to breed with female tortoises from Española Island, which are his closest genetic match, have come to nothing. Other than some far-fetched notions of cloning him, the world has no choice but to sit back and wait for Lonesome George to grow old and die, taking his subspecies into extinction with him. He is a poignant symbol of the destruction that *Homo sapiens* has meted out to the natural world.

The tortoises on Aldabra and the Galápagos are now protected under national law, and the islands on which they live are world heritage sites. We can therefore be reasonably sure (but not certain) that some of the man-made threats of the past, such as over-exploitation, commercial and military development and the introduction of foreign animals, will not be re-enacted. This does not mean that the tortoises, and those campaigning to conserve them, can now rest easy. Far from it.

In November 2000 the staff at the research station of the Charles Darwin Foundation in the Galápagos were forced to abandon their posts as the facility came under violent attack from local fishermen. The scientists and park wardens were forced to spend the night hiding in a mangrove swamp until the Ecuadorian Navy rescued them the next day.

The attack was the culmination of a campaign of intimidation by the fishermen. The station's staff had received death threats and there had been actual isolated physical attacks on scientists and tourists visiting the

Galápagos National Park. The trouble stemmed from tight regulations on the number of lobsters that the fishermen were allowed to catch and sell. The fishermen, who had themselves set the quota at fifty tonnes a year, now wanted it raised. A nervous Ecuadorian government complied with their demand, bringing howls of protest from the conservationists, which, in turn, drove the fishermen to attack them.

After thirty years of harmony, the relationship between the conservationists and the locals is showing severe strain and the research station staff are greatly concerned for the future. 'Obviously, if violence is rewarded, it tends to lead to more violence,' says Alan Tye, the station's acting director.

If the research station should be forced to close by local friction, the tortoises would lose their main source of protection, almost guaranteeing a repetition of the mistakes of the past.

Nor do the problems end there. There are many other dangers present in our world, whose effects are beyond the control of even the most hard-working conservationists. In 2001 an oil tanker was wrecked off the Galápagos, threatening to cover the islands in a thick coating of tar that would have destroyed the marine ecosystem. Nor are all the threats man-made. In 1998 one of Isabela's volcanoes unexpectedly spewed millions of tonnes of liquid lava across the landscape, incinerating everything in its path. The research station used a helicopter to lift twenty tortoises out of the path of the lava. Others were not so lucky and were boiled alive inside their shells.

Then there is the threat of climate change. Whatever the cause, there is little doubt that the earth's climate is warming and the sea level is rising. In 1997 the influx of the rogue El Niño ocean current (whose recent ferocity is commonly blamed on climate change) brought hot sea water into the Galápagos' normally cold-water environment. The result was a change in the local climate and the prolonged disappearance of many local fish and crustaceans, something that only served to annoy the fishermen even more. To quell the trouble the Ecuadorians drafted in more police and park wardens but the atmosphere remains tense.

Tension of a different kind also exists in the Indian Ocean. Aldabra atoll is now safely beyond the reach of most humans and the tortoises there do not face the same types of threat as those in the Galápagos, but the recent changes in the world's climate and sea levels are cause for concern there too. Research on the atoll has found a link between rainfall and tortoise numbers – the wetter it is, the greater the tortoise numbers will be. In recent years there has been a prolonged drought on Aldabra and although the blame cannot be pinned on climate change with certainty, the situation is being monitored closely just in case.

Perhaps more worrying is the slowly rising sea level. On average Aldabra is only a few metres above sea level and it is largely composed of sand and coral rubble, which erodes easily. Even a small rise in the sea level could have an immediate impact, sending the tortoises to a watery grave.

Humanity is often incapable of learning from its past mistakes and it is more through luck than judgement that the magnificent and inspirational animals that we call giant tortoises are still available for us to visit and study in the wild. If, in a hundred years' time, somebody should pick up a battered and faded copy of this book, it is my hope that they too will be able to travel to the Galápagos or Indian Ocean islands and there find the world's giant tortoises, safe and unmolested in their natural environment.

Notes and Sources

GENERAL

CCD   Charles Darwin Correspondence Project, Cambridge University Library, UK
CUL   Cambridge University Library, West Road, Cambridge
NHM   Natural History Museum, Cromwell Road, London
PRO   Public Record Office, Kew, London

PREFACE

This encounter took place on 25 September 1835. The descriptions of the journey are taken from Keynes (2001; pp. 355–7) and Darwin (1839; pp. 488, 492, 498, 503–4). The dialogue and descriptions are based upon Darwin's descriptions of his conversations with Lawson: see Keynes (2001; pp. 355–7).

PART ONE: DISCOVERY

A Sheltered Life, pp. 3–7

For a popular account of recently extinct giant animals, see T. Haines, 2001, *Walking with Beasts*, BBC Books (UK).

places as diverse as . . .: Fossils of giant tortoises are known from all the continents except Australia or Antarctica. See, for example, van Denburgh (1914), pp. 206–7.

In 1497 . . .: The early history of the Mascarene Islands is covered in detail in Grant (1801) and summarised in Stoddart and Peake (1979) although there is a disagreement on dates of discovery between the two.
'. . . turtles of such large dimensions . . .': Grant (1801), p. 22.

The Worthless Islands, pp. 8–14

Berlanga's horrific voyage to the Galápagos is recounted in a detailed letter written by him to the Spanish king. All quotes used come from this letter. See de Berlanga (1884).

For a biography of Berlanga, see his entry in the *Catholic Encyclopaedia* (Thomas Nelson Incorporated; 1976).

'twenty or thirty leagues': In the sixteenth century the length of a league varied between nations. I have taken the Spanish length of 1 league = 4.214 geographical miles (as opposed to the English measurement of 3 geographical miles).

Here Be Giants, pp. 15–22

The history of the early exploration of the Galápagos is covered in Beebe (1924), Günther (1877) and Garman (1917).

'The Spaniards when they first discovered . . .': Dampier (1697), p. 101.

Mauritius, for example, was discovered in 1511 by the Portuguese . . .: Grant (1801); see also *Encyclopaedia Britannica* entry for Mauritius.

'We saw no four-footed creatures . . .': Quote taken from Grant (1801), p. 64. See Leguat (1708) for the original.

Leguat's adventures have been seen as too good to be true by some people and there are many theories that maintain that Leguat's account of his voyage was in fact carefully stitched together from several other people's journals. See Atkinson (1922).

'one of the most delicate morsels . . .': Dubois (1897), p. 80.

John Jourdain: Foster (1905), p. 47.

'better meat for hogs': Herbert (1634), p. 349.

'land tortyses, so great . . .': Herbert (1634), p. 351.

'carry a man with more ease . . .': du Quesne (1887).

another eleven Seychelles atolls: These islands are: African Banks, Agalega, Alphonse, Amirantes, Assumption, Cosmoledo, Denis, Farquhar, Gloriosa, Providence and St Pierre. See Rothschild (1915); Stoddart and Peake (1979).

'great many land-turtle . . .': Stoddart and Peake (1979), p. 154.

PART TWO: INSPIRATION

Just One Species? pp. 25–31

For a biography of Carl Linnaeus, see E. Mayr, 1985, *The Growth of Biological Thought: Diversity, Evolution, and Inheritance*, Harvard University Press (USA).

the specimen was afforded the name . . .: Schweigger (1812). Note: Schweigger's original specimen has been missing for decades, leading to a debate as to which part of the world this tortoise originally came from.

'I was put in the same kind . . .': Delano (1817), p. 376.

'I have seen them . . .': Delano (1817), p. 376.

'The shells of those of James Island . . .': Porter (1822), vol. 1, pp. 214–15.

A Tortoise for Each Island, pp. 32–44

For biographies of Robert FitzRoy, see Mellersh (1968); also P. Nichols, (2003), *Evolution's Captain*, Profile Books (UK).

'wife looked so miserable at the prospect': Mellersh (1968), p. 64.

'biblical idea of Creation': Mellersh (1968), pp. 73–4.

'Although we are mere sojourners . . .': Lyell (1997), p. 102.

'there is a progressive development . . .': Lyell (1997), p. 85.

'no ground for questioning . . .': Lyell (1997), p. 209.

'This voyage is terribly long . . .': Letter from Darwin to William Fox, see *Charles Darwin Calendar* No. 282 (CUL).

'The black rocks heated . . .': Keynes (2001), p. 352.

'The main article . . .': Keynes (2001), p. 356.

Darwin's original notes on the tortoises are to be found in Keynes (2000), pp. 291–3. They were later tidied up and expanded before being published in Darwin (1839), pp. 462–6.

her deck held over thirty adult tortoises . . .: Keynes (2001), p. 353.

'I should think . . .': Keynes (2001), p. 360.

Revelation, pp. 45–71

calling Darwin a blasphemer: Grinnell (1974), pp. 260–1.

'I had been deeply impressed . . .': Barlow (1958), pp. 118–19.

'I last wrote to you . . .': Keynes (2001), p. 357.

'The [mockingbird] specimens from Chatham . . .': Keynes (2001), p. 357; Sulloway (1982b), p. 328. See Sulloway's notes on the dating of this comment.

'There is no other animal . . .': FitzRoy (1839), Chapter 21.

Amid the qualified observations . . .: Darwin (1839, 1845), see appendices.

. . . with John Gray, the resident reptile expert . . .: Darwin (1845), p. 394; see Günther (1877), p. 6.

Testudo indica: Gray (1831).

the museum's own collections: See Günther (1877) for a list of the specimens held in the British Museum.

'series of ground finches . . .': Paraphrased from Sulloway (1982b), p. 358.

'I cannot enter into any further details . . .': Sulloway (1982b), p. 360.

'If they are varieties . . .': Sulloway (1982b), p. 379.

'such facts would undermine . . .': Keynes (2001), p. 357.

'Mr Lawson states he can . . .': Keynes (2000), p. 291.

'I did not for some . . .': Darwin (1845), p. 394.

Waterhouse was a busy man . . .: See CUL DAR 181: 18, 181: 17, 181: 19.

a letter to his Cambridge mentor John Henslow . . .: Sulloway (1982b), p. 370.

'doves and finches swarmed around its margin': Sulloway (1982a), p. 19.

This included thirteen finch specimens . . .: Sulloway 1982b.

'Opened the first notebook . . .': de Beer (1960), p. 23.

'It was confidently asserted . . .': Darwin (1839), pp. 455–6.

Testudo elephantina, *Testudo daudinii* and *Testudo peltastes*: Duméril and Bibron (1835), pp. 80, 110, 123.

The two men talked . . .: Duméril and Bibron (1835), p. 80.

'. . . at least two distinct species of tortoise . . .': Darwin (1839), p. 628; Bibron also told Darwin that there were at least two species of marine iguana from the Galápagos.

'The French Bibron . . .': CUL DAR 118: 17.

'There is every reason . . .': Darwin (1839), pp. 628–9.

Darwin watched as his old mentor . . .: Desmond and Moore (1992), p. 274.

'there are several instances . . .': Darwin (1839), p. 628.

'flora of this archipelago . . .': Darwin (1887), vol. 2, p. 20.

'I cannot tell you how delighted . . .': Darwin (1887), vol. 2, p. 22.

'I have not as yet noted . . .': Darwin (1845), Chapter 17.

'Reviewing the facts here given . . .': Darwin (1845), Chapter 17.

Ironies, pp. 72–5

David Porter, a captain in the American Navy: See Porter (1822).

At least one ornithologist . . .: Swarth (1931).

PART THREE: DECIMATION

The Indian Ocean Tortoises, pp. 79–84

[not once does Darwin] mention . . .: Keynes (2001), pp. 419–27.

'Grey parrots are also common there . . .': Hachisuka (1953), p. 64.

'reputed more for wonder . . .': Fuller (1988), p. 122.

To add insult, the bird . . .: Grant (1801), p. 25.

For a full account of the dodo's extinction, see Hachisuka (1953) or Fuller (1988).

In 1673, Hubert Hugo recounts . . .: Stoddart and Peake (1979), p. 420.

The trade continued for several years . . .: Dupon (1969), pp. 24–40.

the last report of a wild tortoise was made in 1778 . . .: For last references to the Mascarene tortoises, see Rothschild (1915); Stoddart and Peake (1979); Dupon (1969).

Available government records . . . : For an account of the trade in Seychelles tortoises, see Fauvel (1900).

'private vessels have carried them off . . .': Rothschild (1915), p. 421.

In 1787 it was estimated . . .: Fauvel (1900).

The Safe Haven, pp. 85–92

'exceed in sweetness . . .': Dampier (1697).

For the Penguin Island incident, see PRO BT6/95, pp. 227–79.

For the history of the southern whaling fleet, see Harmer (1928, 1931) and Jackson (1978).

'If that claim were given': Hansard (1816), vol. 28, column 770.

Details of Colnett's life and career can be found in Colnett (1798).

'during [the] interval, death had deprived me . . .': Colnett (1798), p. vi.

We know that Colnett was familiar with Dampier's work because he continually quotes him in his own observations: see Colnett (1798).

'The Spaniards are said . . .': Colnett (1798), p. 60.

'So disgusting is their appearance . . .': Colnett (1798), p. 56.

'an excellent broth': Colnett (1798), p. 56.

'roused the hideous animal . . .': Colnett (1798), p. 134.

'The inhospitable appearance of this place . . .': Colnett (1798), p. 143.

NOTES AND SOURCES

'These isles . . .': Colnett (1798), p. 145.

'all the luxuries of the sea . . .': Colnett (1798), p. 159.

The Arrival of the Whaling Fleet, pp. 93–105

'Post Office Bay': Beebe (1924), p. 370.

'I have had these animals . . .': Townsend (1925).

'Shortly after the ship *Niger* . . .': Townsend (1925).

'Presently to my surprise . . .': This description is built from passages in three chapters of Davis (1874).

'Left bread and water . . .': Townsend (1925).

Each whaling boat would load . . .: Based on figures in Townsend (1925).

The first populations to suffer . . .: Based on figures in Townsend (1925).

. . . but Charles Townsend . . .: Townsend (1925).

'In 1801 the Pacific sailors . . .': Jackson (1978), p. 135.

'Nothing, perhaps, can be more disagreeable . . .': Porter (1822), Chapter VI.

For details of Porter's life, see Porter (1822).

'The [turpining] parties . . .': Edited from Porter (1822), Chapter X.

'The most of those we took on board . . .': Porter (1822), Chapter IX.

'leave it to those . . .': Edited from Porter (1822), Chapter IX.

According to Charles Townsend's figures . . .: Townsend (1925).

Settlers, pp. 106–12

'The appearance of this man . . .': Porter (1822), Chapter VI.

'I have made repeated applications . . .': Porter (1822), Chapter VI.

Patrick Watkins: The only expanded description of the eccentric Watkins comes from Porter (1822), Chapter VI.

'The inhabitants here . . .': Keynes (2001), p. 356.

'Future navigators may perhaps . . .': Porter (1822), Chapter IX.

Around 1840 the Santa María Island tortoise became extinct . . .: The last record of a ship taking tortoises from Santa María Island was in 1847 but it is thought that these animals were probably transferred there from other islands. From Townsend (1925).

San Salvador and Santa Fe islands . . .: The last recorded tortoise on Santa Fe was taken by the ship *Henry N. Crapo* in 1853. From Townsend (1925).

By the turn of the twentieth century . . .: For a history of Galápagos colonisation, see Beebe (1924).

PART FOUR: OBSESSION

La Mare aux Songes, pp. 115–17

'I repaired to this spot . . .': *Philosophical Transactions of the Royal Society*, (1875) 165: p. 251.

A Champion for the Tortoises, pp. 118–34

For the life of Albert Günther, see Günther (1975a) and the Günther archive (NHM).

'been scattered by man': Günther archive (NHM), Box 19, p. 1.

'examine specimens from [all] islands . . .': Günther archive (NHM), Box 19, p. 1.

'If you keep my Aldabra tortoise . . .': Günther archive (NHM), Box 19, p. 1; Günther (1975a), p. 318.

'By the beginning of our century . . .': Günther (1898).

'I hear that the Mauritius government . . .': Günther archive (NHM), Box 19, p. 3.

'The place for forming a park . . .': Günther archive (NHM), Box 19, p. 4.

'I have taken every precaution . . .': Günther archive (NHM), Box 19, p. 6.

'I would have much pleasure . . .': Günther archive (NHM), Box 19, p. 5.

'it is perfectly monstrous . . .': Günther archive (NHM), Box 19, Günther (1975a), p. 319.

'To His Excellency . . .': Günther archive (NHM), Box 19, pp. 22–5.

'How much,' asked the governor . . .: Günther archive (NHM), Box 19, p. 26.

'[You] are not correct . . .': Günther archive (NHM), Box 19, pp. 31–2.

'it is probable that even . . .': Günther archive (NHM), Box 19, pp. 31–2.

once owned by Queen Victoria . . .: *The Times* (UK), 26 July 1850.

and offered to obtain . . .: Günther archive (NHM), Box 19, p. 34.

the central course of the wedding feast . . .: Günther archive (NHM), Box 19, pp. 30, 56; Günther (1975a), p. 318.

'Shipment of Two Tortoises . . .': Günther archive (NHM), Box 20, p. 6.

'the understanding that the Trustees . . .': Günther archive (NHM), Box 20, p. 7.

'in case of difficulty arising . . .': Günther archive (NHM), Box 20, p. 7.

Günther was not in the Museum's good books: Günther archive (NHM), Box 19, p. 44.

reluctantly Günther paid the tax . . .: Günther archive (NHM), Box 19, p. 56; Box 20, pp. 6–7.

A deal had been struck . . .: Günther archive (NHM), Box 20, pp. 6–7.

some of the covert transactions . . .: For example, see the purchase of the first *Archaeopteryx* specimen in my book *Bones of Contention*.

'most cherished of boyhood memories . . .': Günther archive (NHM), Box 20, p. 7.

an instant tourist attraction . . .: *The Times* (UK), 20 March 1875, p. 9.

'You may have a look . . .': *The Living Age* (USA), 127: 1636, pp. 190–2, 16 October 1875.

Had it been left alone in its owner's . . .: Günther (1877), p. 1.

Understanding the Tortoises, pp. 135–43

get his hands on over forty giant tortoise specimens . . .: For a list of these specimens, see Günther (1877).

'Under the name *Testudo indica* . . .': Günther (1875), p. 252.

'How,' asked Günther . . .': The quotes come from a meeting at which Günther read out his 1877 paper; see Günther (1877), p. 9.

'The giant tortoises . . .': Günther (1877), p. 9, paraphrased by me.

'Last night we were party . . .': Letter from Hooker to Darwin, 27 January 1877, Darwin archive, Cambridge University.

'How the deuce they got . . .': Letter from Darwin to Hooker, 28 January 1877, Darwin archive, Cambridge University.

Alexander Agassiz visited the islands . . .: See *Bulletin of the Museum of Comparative Zoology, Harvard College*, 23: 1, 1892.

George Baur, a fellow scientist . . .: See *American Naturalist*, March–April 1891.

deliberately misquoting their papers . . .: See *Science* (1892) 19: 477, p. 176.

bodily picked up from South America . . .: Beebe (1924), p. 420.

Save the Tortoises! pp. 144–8

'there is plenty of good ground . . .': Günther archive (NHM), Box 20, p. 9; Günther (1975a), p. 425.

'Many vessels call at Aldabra . . .': Letter dated 21 February 1879, quoted in Günther archive (NHM), Box 20, p. 9.

'the tortoises never quarrel among themselves': Günther archive (NHM), Box 19, p. 83; see also Stoddart and Peake, 1979.

'Mr Button of the Seychelles . . .': Günther archive (NHM), Box 19, p. 90. Note: Despite his interest in giant tortoises, Colonel Gordon grew bored with his sabbatical in Palestine and pleaded for another posting. He was sent to the Sudan where, on 26 January 1885, he was murdered on the palace steps in Khartoum by marauding local troops. The relief soldiers, for which Gordon had been pleading for months, arrived two days later.

'Having been informed that Aldabra . . .': Günther archive (NHM), Box 19, p. 90.

'fail to bear in mind . . .': Günther archive (NHM), Box 19, p. 98.

'completely at liberty . . .': Nature, 23 (1883), p. 398.

'thriving, if not multiplying': *The Daily Graphic* (UK), 18 January 1893, p. 14.

'I am sorry . . .': *Nature*, 23 (1883), p. 398.

The Final Showdown, pp. 149–53

'unlikely to kill the goose . . .': Günther archive (NHM), Box 20, p. 10.

'When Mr Spurs first went . . .': Günther archive (NHM), Box 19, p. 106.

'The information that I had gathered . . .': Günther archive (NHM), Box 19, no page number.

'[The Aldabra Islands] are now leased by the Mauritian Government . . .': *The Times* (UK), 12 January 1893, p. 10.

As a Liberal MP . . .: Günther archive (NHM), Box 19, p. 139.

'Mr Spurs' estimate . . .' : Günther archive (NHM), Box 19, p. 146.

'that the killing of the tortoises . . .': Letter 19 May 1899, Günther archive (NHM), Box 20, p. 14.

Galápagos or Bust, pp. 154–67

Günther was quite taken aback . . .: Günther's early dealings with Walter Rothschild are dealt with in Günther, 1975a, pp. 417–32, and Rothschild, 1983, pp. 57, 65, 72 and 91.

Walter Rothschild's life: Walter Rothschild was notoriously secretive and prior to his death destroyed almost all his personal letters and papers, making it very difficult for historians to reconstruct his life. Much of the information on Walter's family life for this chapter comes from the recollections of his niece Miriam. See Rothschild (1983).

write more than 320 others . . .: Günther (1975a), p. 419; Rothschild (1983), p. 327.

'[Walter] sought to cultivate my friendship . . .': Rothschild (1983), p. 72.

some 38,000 pinned butterflies . . .: Günther (1975a), p. 418.

over 300 collecting expeditions . . . .: Rothschild (1983), p. 152.

Walter had to make do . . .: Günther (1975a), p. 426.

'every tortoise they could . . .': Letter from Dr Harris to Günther, 8 February 1897, Günther (1975a), p. 426.

'Your proposals . . .': Rothschild (1983), p. 185.

'My chief reason . . .': Letter from Günther to Rothschild, 27 December 1897, Günther archive (NHM), Box 20.

News of [Bullock's] death . . .: Rothschild (1983), pp. 185–96.

'If I was his father . . .': Rothschild (1983), p. 190.

'Don't let this fall through . . .': Rothschild (1983), p. 189.

'the trip will cost . . .': Rothschild (1983), p. 193.

'After a long walk . . .': van Denburgh (1914), pp. 229–30.

'if all's well . . .': Letter from Rothschild to Günther, 27 December 1897, Günther (1975a), pp. 426–7.

'I now have fifty-five . . .': Günther archive (NHM), Box 20. Five of the tortoises must have died in transit. The tortoises initially went to Rothschild's Paris chateau before being moved to Tring.

there were over 144 live tortoises . . . .: Rothschild (1983), p. 204.

Walter purchased Rotumah . . .: Günther archive (NHM), Box 20; Rothschild (1983), pp. 202–3; Scott Thomson, personal communication to author.

no captive giant tortoise . . . .: Rothschild (1915, p. 404) notes that a tortoise collected in 1900 'has lived fifteen years in England – a record, I believe, for any Giant Tortoise'.

The Errant Playboy, pp. 168–72

'to tell Boulanger . . .': Rothschild (1983), p. 203; Günther (1975a), p. 428.

'If it is not bothering . . .': Günther (1975a), p. 419.

started to blackmail [Walter] . . . : Details of the blackmail are scanty. Miriam Rothschild (1983) mentions it many times in her biography of Walter and it has apparently been the subject of other articles about Lord Rothschild.

'The inevitable kindness . . .': Letter to Robert Günther, 2 February 1914, quoted in Günther (1975a), p. 429.

'loss to British ornithology': *The Times* (UK), 11 March 1931.

On cataloguing . . . : Rothschild (1983), p. 2.

PART FIVE: PETS

The Well-Travelled Tortoises, pp. 175–6

'is ascertained to have . . .': *The Times* (UK), 7 November 1810.

The stuffed corpse . . . : *The Times* (UK), 30 May 1966, p. 10.

Darwin's Missing Tortoise, pp. 177–86

'Tortoise Recalled': *Courier Mail* (Brisbane), 7 August 1994.

The history of Scott Thomson's research into Harriet was told to me by Dr Thomson and staff at the Australia Zoo. A summary can be found in Thomson *et al.* (1996).

David Fleay's life is told through a series of his books. For example, see Fleay (1944, 1960).

Harry, therefore, became Harriet: See Fleay (1960), pp. 44–5.

find evidence that she was in . . . : Scott Thomson, personal communication to the author.

Thomson contacted Ed Loveday . . . : Scott Thomson, personal communication to the author.

Tom the tortoise: Although not mentioned in the text, Thomson managed to track down another of the Botanical Gardens' tortoises. This animal, called Tom, died in 1929 and is preserved in the Brisbane Museum. It turned out to be from San Cristóbal Islands in the Galápagos. See Thomson *et al.* (1996).

Digging into Harriet's Past, pp. 187–97

Scott Thomson's paper: Thomson *et al.* (1996).

a mighty flood . . .: See Anon., *Souvenir of Floods: Southern Queensland, February, 1893*, Offices of *The Telegraph* and *The Week*, Brisbane, 12pp, 1893.

On 18 September 1835 . . . : FitzRoy (1839), p. 488.

on 12 October . . . : FitzRoy (1839), p. 498.

'On board the *Beagle* . . .': FitzRoy (1839), p. 504.

'young ones and . . .': Darwin (1839), p. 394.

The precise fate . . .: Desmond and Moore (1992), p. 697; interview with Nino Strachey given to www.aboutdarwin.com

'Covington's little tortoise . . .': CUL DAR, 29.3: 40, MS p. 7v.

reptile specimens donated to the British Museum . . .: NHM (personal communication to the author).

British Museum's Zoological Accessions book . . .: NHM Zoological Accessions, 1837, p. 1 – NHM DF218/1, LIS-Archives. These two tortoises possibly appear in the Günther (1877, p. 70) résumé as being *T. nigrata* although there is now no sign of them in the Museum (NHM, personal communication to the author).

'a very small [tortoise] lived . . .': FitzRoy (1839), p. 504.

'that I brought from three islands': Darwin (1845), p. 394.

For Darwin's life after the *Beagle*, see Desmond and Moore (1992); also *The Correspondence of Charles Darwin*, Cambridge University Press, vols. 1 and 2 (1985, 1986).

Covington left Darwin's employment . . . : CCD, 513 to 515.

living in a tall London brick house . . . : CUL DAR 204: 156, 158.

Darwin's health: Darwin suffered from poor health for much of his adult life. Some of this may well have been psychological in nature. In his autobiography he writes of heart palpitations prior to the *Beagle* voyage, which, even then, he believed to be a symptom of heart disease.

no specimens in their collections . . .: I have contacted curators at all the named institutions to ask if they have any specimens that could match the two missing *Beagle* tortoises. None had any record of having received them.

For Wickham's navy career, see *Biographical Dictionary of the Officers of the Royal Navy*, published *circa* 1847. The (battered) copy I found was in the Institute for Historical Research, London.

For the 1837–43 *Beagle* logbooks, see National Maritime Museum 1 MRF/123. Other records for the voyage also exist there.

Moreton Bay in 1841 . . .: Craig (1925).

Hannibal pulled some strings . . .: See *Newstead House: The History*, Friends of Newstead House (1996), p. 15.

March 1841 census: On 2 March 1841 there is a John Wickham listed in Lane Cove Wharf, Parish Gordon, County Cumberland, Sydney. Return 59, p. 159, reel 2223.

a letter written by Lord Stanley . . .: Historical Research and Access (Australia); *Governor's Dispatches*, April 1842 to June 1843, pp. 48 and 776.

I have searched for evidence . . .: Sites searched include the Society of Genealogists' library, the Institute for Historical Research, the National Maritime Museum, the British Library and the Public Record Office.

Darwin and Wickham: There is no record of any correspondence between Darwin and Wickham in the aftermath of the *Beagle* voyage although the two did meet again in 1862 at a minor ship's reunion organised by Bartholomew Sullivan. CUL DAR 177: 275, 276, 277; 171: 146.

known to survive is fifteen years . . .: Rothschild (1915), p. 404.

The Fate of the *Beagle* Tortoises, pp. 199–205

Günther and Darwin: see Günther (1975a), pp. 453–70.

Thomas Huxley landed the first blow . . .: See my book *Bones of Contention*, Chapter 2.

'You have my most . . .': Günther (1975b), p. 30.

'I find that I did not bring home . . .': Günther (1975b). Note: Darwin mentions that the surgeon, Benjamin Bynoe, brought back some tortoises, which suggests that there may have been more than just the four individuals that we know of. I have found no references to Bynoe in connection with the tortoises.

Royal United Services Museum: Information on the Museum and its archive came from *The Times* (index 1795–1905) and from the curator of the Army Museum. Due to pressure of time my searches of the Army Museum's archive were not exhaustive and it is possible that a mention of the tortoises exists there somewhere.

Harriet's DNA test: Scott Davis and Scott Thomson, personal communication to the author. See also Thomson *et al.* (1996).

'My hunch is that . . .': Scott Thomson, personal communication to the author.

'It does not seem likely . . .': Frank Sulloway, personal communication to the author.

'Harriet's nuclear DNA . . .': Scott Davis, personal communication to the author.

Townsend lists only . . .: Townsend (1925).

Marion's Tortoise, pp. 206–10

a remaining population . . .: Merlen (1999), p. 8.

'only spot in the Indian Ocean . . .': Günther (1877), p. 3.

'it is just possible . . .': *The Illustrated London News*, 3 December 1893, p. 715.

The French pleaded . . .: Letter from Pasfield-Oliver to Günther, 7 September 1893; Günther archive, Box 19, p. 118; *The Times* (UK), 27 January 1930.

commanding officer would sit . . .: *The Times* (UK), 29 January 1930.

'tortoise to be the Rodrigues species . . .': *The Times* (UK), 11 January 1893.

even quoting from a letter . . .: *The Times* (UK), 11 January 1893.

'If I had known . . .': *The Times* (UK), 12 January 1893.

'I fear since they . . .': Letter to Günther, 1 December 1893; Günther archive (NHM), Box 19.

One soldier later recalled . . .: *The Times* (UK), 29 January 1930.

Its corpse was found . . .: *The Times* (UK), 27 January 1930.

flooded with letters . . .: *The Times* (UK), 29 and 30 January, 5 and 6 February 1930.

PART SIX: RECOVERY

In the Name of Science, pp. 213–26

'These creatures are so nearly . . .': van Denburgh (1914), p. 237.

organised by Stanford University . . .: For details of this expedition, see Heller (1903).

'They are very determined travellers . . .': Heller (1903).

another expedition . . .: Noyes' expedition is briefly described in van Denburgh (1914), p. 237.

'At the rate of destruction . . .': Beck (1903).

'The outfit . . .': Beck (1903).

'It is only within . . .': Beck (1903).

Although there is some doubt . . .: See Pritchard (1996).

'After seeing on this mountain . . .': Beck (1903).

the true number of tortoises . . .: Townsend (1925); Merlen (1999), p. 3.

'The only remaining hope . . .': Townsend (1925).

the last great scientific expedition . . .: For details of the New York Zoological Society's 1930 expedition, see Townsend (1931).

The threat, he argued . . .: Townsend (1942).

Aldabra's Legacy, pp. 227–39

Scientists blamed the shortfall . . .: The Nature Protection Trust of the Seychelles, personal communication to the author.

If Günther's colony did not disappear . . .: See references in Stoddart and Peake (1979).

Amid a highly detailed breakdown . . .: [*Seychelles Crown Land Department Report on the Aldabra Group of Islands*, 1895].

William Louis Abbott . . . .: Abbott (1893).

a succession of visitors . . . .: Fryer (1911).

'no plan will effectively prevent . . .': Davidson (1911).

'it would be possible to live . . .': Fryer (1911).

'still to be found . . .': Dupont (1907).

J. Fryer was more specific . . . .: Fryer (1910).

The Seychelles Guano Company Limited: PRO CO 530/108; PRO CO 1036/135–6.

The Seychelles Guano Company and malaria: PRO CO 503/223.

Records indicate . . . .: Dupont (1929); Stoddart and Peake (1979).

'The atoll is unique . . .': Stoddart (1968).

In a report to the scientific journal *Nature* . . . .: Stoddart and Wright (1967).

'We have assured the Royal Society . . .': Hansard, vol. 745, columns 253–4.

Tortoise expert W. Bourne . . . .: *Science*, 29 September 1967, p. 1511.

Among the eminent names . . . .: *The Times* (UK), 16 August 1967; *New Scientist*, 24 August 1967.

Tam Dalyell: see PRO PREM 13/680.

The Royal Society commented . . . .: Stoddart (1967), p. 67.

The *Economist* came out . . . .: *The Economist* (UK), 11 October 1967.

For details of the debates over Aldabra in the UK parliament, see Hansard, vol. 755, columns 26, 68, 114, 126–35, 156.

For negotiations between the UK and US governments over the base, see PRO FCO 83/7; 32/113–16.

founded a permanent research station . . . .: PRO FCO 32/126–7; 26/450.

A detailed census . . . .: Bourn and Coe (1978).

the dispossessed Diego Garcian people . . .: *Time* (USA), 20 February 1998; BBC News, 9 October 2003.

a follow-up to the 1973 tortoise census . . .: Bourn *et al.* (1999).

only thirty were left . . .: Bourn *et al.* (1999).

The Charles Darwin Foundation, pp. 240–48

Sources for this chapter include: Charles Darwin Foundation (personal communication to the author); Cayot *et al.* (1994); MacFarland (1972); Pritchard (1996); Benchley (1999).

'To a geologist . . .': *Washington Post*, 3 April 1944.

PART SEVEN: IN THE BLOOD

Lumpers and Splitters, pp. 251–61

The last great surveys . . .: Rothschild (1915), pp. 418–19.

The Galápagos tortoises and *Geochelone*: See Loveridge and Williams (1957).

Nicholas Arnold's new classification: Arnold (1979).

Dipsochelys and Bour's species revision: See Bour (1982, 1984).

Gerlach and Canning's revision: Gerlach and Canning (1998a, 1998b).

Two new Seychelles species: Nature Protection Trust for the Seychelles, personal communication to the author; Gerlach and Canning (1998a, 1998b).

The 1997 genetic fingerprinting: Gerlach and Canning (1998b).

In 2002 another group of scientists: Palkovacs *et al.* (2003).

The answer was no: Austin *et al.* (2003).

the genetics of the Mascarene tortoises: Austin and Arnold (2001).

the results being published between 1999 and 2003: Louis (1999); Caccone *et al.* (1999); Ciofi *et al.* (2002); Beheregaray *et al.* (2003).

all eleven of the living species and analysed: Caccone *et al.* (1999).

The Tortoises' Origins, pp. 262–9

the oldest division into subspecies: Caccone *et al.* (1999).

between about 5 and 2 million years ago: Louis (1999); Caccone *et al.* (1999); Ciofi *et al.* (2002); Beheregaray *et al.* (2003).

The newly arrived *Cylindraspis inepta*: Austin and Arnold (2001); Caccone *et al.* (1999).

Others said that they had been swept: Bour (1982, 1984).

high sea level during the last ice age: Palkovacs *et al.* (2002, 2003); Austin *et al.* (2003).

Epilogue

Sources include the Charles Darwin Foundation (personal communication to the author); *New Scientist* (4 December 2000); BBC News Archive (8 September 1998, 6 October 1998, 9 November 1999, 3 October 2003); and Benchley (1999).

Bibliography

This bibliography refers chiefly to the papers cited in the Notes and Sources and does not represent the full list of publications consulted during the preparation of this book.

Abbott, W.L., 1893, 'Notes on the natural history of Aldabra, Assumption and Glorioso Islands, Indian Ocean', *Proceedings of the US National Museum*, 16, pp. 759–64.

Arnold, E.N., 1979, 'Indian Ocean giant tortoises: their systematics and island adaptions', *Philosophical Transactions of the Royal Society of London*, Series B, 286, pp. 125–45.

Atkinson, G., 1922, *The Extraordinary Voyage in French Literature from 1700 to 1720*, Librairie Ancienne Honoré Champion (France).

Austin, J.J. and Arnold, E.N., 2001, 'Ancient mitochondrial DNA and morphology elucidate an extinct island radiation of Indian Ocean giant tortoises (*Cylindraspis*)', *Philosophical Transactions of the Royal Society of London*, Series B, 268, pp. 2515–23.

Austin, J.J., Arnold, E.N. and Bour, R., 2003, 'Was there a second adaptive radiation of giant tortoises in the Indian Ocean? Using mitochondrial DNA to investigate speciation and biogeography of *Aldabrachelys*', *Molecular Ecology*, 12, pp. 1415–24.

Barlow, N., 1958, *Charles Darwin's Autobiography with Original Omissions Removed*, Collins (UK).

Baur, G., 1889, 'The gigantic land tortoises of the Galápagos Islands', *American Naturalist*, 23: 276, pp. 1039–57.

Beck, R.H., 1903, 'In the home of the giant tortoise', *Seventh Annual Report of the New York Zoological Society*, 7, pp. 160–74.

Beebe, W., 1924, *Galápagos: World's End*, Putnam & Sons (USA).

Beheregaray, L.B., Ciofi, C., Caccone, A., Gibbs, J. and Powell, J.R., 2003, 'Genetic divergence, phylogeography and conservation units of giant tortoises from Santa Cruz and Pinzón, Galápagos Islands', *Conservation Genetics*, 4, pp. 31–46.

Behler, J.L., 1974, 'Vanishing Species: Tortoises – Part 1', *Animal Kingdom*, 77: 2, p. 33.

Benchley, P., 1999, 'Galápagos: paradise in peril', *National Geographic*, 195, p. 4.

Bour, R., 1982, 'Contribution à la connaissance des Tortues terrestres des Seychelles: definition du genre endémique et description d'une espèce nouvelle probablement originaire des îles granitiques et au bord de l'extinction', *Comptes Rendus de l'Académie des Sciences, Paris*, Series 3, 295, pp. 117–22.

Bour, R., 1984, 'Taxonomy, history and geography of Seychelles land tortoises and freshwater turtles', in *Biogeography and Ecology of the Seychelles Islands* (ed. Stoddart, D.R.), W. Junk Publishers (The Hague), pp. 281–307.

Bourn, D. and Coe, M., 1978, 'The size, structure and distribution of the giant tortoise population on Aldabra', *Philosophical Transactions of the Royal Society of London*, Series B, 282, pp. 139–78.

Bourn, D., Gibson, C., Augeri, D., Wilson, C.J., Church, J. and Hay, S.I., 1999, 'The rise and fall of the Aldabran giant tortoise population', *Philosophical Transactions of the Royal Society of London*, Series B, 266, pp. 1091–1100.

Caccone, A., Gibbs, J.P., Ketmaier, V., Suatoni, E. and Powell, J.R., 1999, 'Origin and evolutionary relationships of giant Galápagos tortoises', *Proceedings of the National Academy of Sciences of the USA*, 96, pp. 13223–8.

Cayot, L.J., Snell, H.L., Llerena, W., and Snell, H.M., 1994, 'Conservation biology of Galápagos reptiles: twenty-five years of successful research and management', in *Captive Management and Conservation of Amphibians and Reptiles* (eds. Murphy, J. and Collins, J.C.), SSAR (USA), pp. 297–305.

Ciofi, C., Milinkovitch, M.C., Gibbs, J.P., Caccone, A. and Powell, J.R., 2002, 'Microsatellite analysis of genetic divergence among populations of giant Galápagos tortoises', *Molecular Ecology*, 11, pp. 2265–83.

Colnett, J., 1798, *A Voyage to the South Atlantic*, reprinted in 1968 by Da Capo Press (USA).

Craig, W.W., 1925, *Moreton Bay Settlement or Queensland before Separation, 1770–1859*, Watson, Ferguson & Co. (Australia).

Dampier, W., 1697, *A New Voyage Round the World*, reprinted in 1919 by Dover Publications (USA).

Darwin, C.R., 1839, *Journal of Researches into the Geology and Natural History of the Various Countries Visited by H.M.S. Beagle, under the Command of Captain Robert FitzRoy R.N., from 1832 to 1836*, Henry Colburn (UK).

Darwin, C.R., 1845, *Journal of Researches into the Geology and Natural History of the Various Countries Visited by H.M.S.* Beagle, John Murray (UK), second edition.

Darwin, C.R., 1989, *Voyage of the Beagle*, Penguin (UK), abridged version.

Darwin, F., 1887, *The Life and Letters of Charles Darwin*, 3 vols., John Murray (UK).

Davidson, W.E., 1911, 'Land Tortoises of the Seychelles', *Proceedings of the Royal Society of London*, 1911, pp. 622–4.

Davis, W.M., 1874, *The Nimrod of the Sea*, reprinted in 1924 by Charles E. Lauriat Co. (USA).

de Beer, G., 1960, 'Darwin's notebooks on transmutation of species', *Bulletin of the British Museum (Natural History)*, Series 2, pp. 23–83.

de Berlanga, T., 1884, 'A letter to his majesty from Fray Tomás de Berlanga, describing his voyage from Panama to Puerto Viejo, and the hardships he encountered in this navigation', in *Colección de Documentos Inéditos Relativos al Descubrimiento, Conquista y Organización de las Antiguas Posesiones Españolas de América y Oceania*, vol. 41, Manuel G. Hernández (Madrid), pp. 538–44.

Delano, A., 1817, *A Narrative of Voyages and Travels*, E.G. House (USA).

Desmond, A. and Moore, J., 1992, *Darwin*, Penguin (UK).

Dubois, S., 1897, *The Voyages made by Sieur Dubois*, Nutt (UK).

Duméril, A.M.C. and Bibron, G., 1834, *Erpétologie Générale ou Histoire Naturelle Complète des Reptiles*, Librairie Encylopédique de Roret (France).

Dupon, J.F., 1969, *Recueil de Documents pour servir à l'histoire de Rodrigues*, Coquet (Mauritius).

Dupont, R.P., 1907, *Report on a Visit of Investigation to St. Pierre, Astove, Cosmoledo, Assumption and the Aldabra Group*, Government Printers (Seychelles).

Dupont, R.P., 1929, *Report on a Visit of Investigation to the Principal Outlying Islands of the Seychelles Archipelago*, Government Printers (Seychelles).

du Quesne, H., 1887, *Projet de république à l'Ile d'Eden*, Thomas Sauzier (Paris).

Fauvel, A.A., 1900, 'Textes inédites concernant les tortues de terre gigantesques', *Bulletin de la Musée d'Histoire Naturelle, Paris*, 6, pp. 170–4.

FitzRoy, R., 1839, *Narrative of the Surveying Voyages of His Majesty's Ships* Adventure and Beagle *between the Years 1826 and 1836*, Henry Colburn (UK).

Fleay, D.H., 1944, *We Breed the Platypus*, Robert & Mullens (Australia).

Fleay, D.H., 1960, *Living with Animals*, Lansdowne Press (Australia).

Foster, W., 1905, *The Journal of John Jourdain 1608–1617*, Hakluyt Society (UK).

Fryer, J.C.F., 1910, 'The south-west Indian Ocean', *Geographical Journal*, 37, pp. 249–68.

Fryer, J.C.F., 1911, 'The structure and formation of Aldabra and neighbouring islands', *Transactions of the Linnaean Society of London*, Zoology Series, 14, pp. 397–442.

Fuller, E., 1988, *Extinct Birds*, Facts on File (USA).

Garman, S., 1917, 'The Galápagos tortoises', *Memoirs of the Museum of Comparative Zoology*, 30: 4, pp. 261–96.

Gerlach, J. and Canning, K.L., 1998a, 'Taxonomy of Indian Ocean Tortoises (*Dipsochelys*)', *Chelonian Conservation and Biology*, 3, pp. 3–19.

Gerlach, J. and Canning, K.L., 1998b, 'Identification of Seychelles giant tortoises', *Chelonian Conservation and Biology*, 3, pp. 133–5.

Grant, C., 1801, *The History of Mauritius*, reprinted in 1995 by Asian Educational Services (India).

Gray, J.E., 1831, *Synopsis Reptilium, Part One: Cataphracta. Treuttel*, Wurtz (London).

Grinnell, G., 1974, 'The rise and fall of Darwin's first theory of evolution', *Journal of the History of Biology*, 7, pp. 259–73.

Günther, A., 1875, 'Description of the living and extinct races of gigantic land-tortoises. Part 1–2. Introduction and the tortoises of the Galápagos Islands', *Philosophical Transactions of the Royal Society of London*, Biological Sciences, 165, pp. 251–84.

Günther, A., 1877, *The Gigantic Land-Tortoises (Living and Extinct) in the Collection of the British Museum*, Taylor & Francis (London).

Günther, A., 1896, 'Notes on *Testudo ephippium*, Günther', *Novitates Zoologicae*, 3, p. 329.

Günther, A., 1898, 'Recent progress on our knowledge of the distribution of the gigantic land tortoises', *Proceedings of the Linnaean Society of London*, 110, pp. 14–29.

Günther, A.E., 1975a, *A Century of Zoology*, Dawsons (UK).

Günther, A.E., 1975b, 'The Darwin letters at Shrewsbury School', *Notes and Records of the Royal Society of London*, 30, pp. 25–43.

Hachisuka, M., 1953, *The Dodo and Kindred Birds*, Witherby (UK).

Harmer, S.F., 1928, 'History of whaling', *Proceedings of the Linnaean Society of London*, 140, pp. 51–95.

Harmer, S.F., 1931, 'Southern whaling', *Proceedings of the Linnaean Society of London*, 142, pp. 85–163.

Heller, E., 1903, 'Papers from the Hopkins–Stanford Galápagos expedition, 1898–99. 14. Reptiles', *Proceedings of the Washington Academy of Sciences*, 5, pp. 39–98.

Herbert, T., 1634, *A Relation of Some Years Travail, begunne Anno 1626*, Stansby & Bloome (UK).

Jackson, J., 1978, *The British Whaling Trade*, Adam & Charles Black (UK).

Keynes, R.D., 2000, *Charles Darwin's Zoology Notes and Specimen Lists from H.M.S. Beagle*, Cambridge University Press (UK).

Keynes, R.D., 2001, *Charles Darwin's Beagle Diary*, Cambridge University Press (UK).

Leguat, F., 1708, *A New Voyage to the East Indies*, Bonwicke (UK).

Louis, E., 1999, 'Phylogenetic relationships of the Galápagos tortoise *Geochelone nigra* from microsatellite and mitochondrial DNA data', *Proceedings of the American Association for the Advancement of Science, Pacific Division*, 18: 1, pp. 64–5.

Loveridge, A. and Williams, E.E., 1957, 'Revision of the African tortoises and the turtles of the suborder *Cryptodira*', *Bulletin of the Museum of Comparative Zoology, Harvard*, 115, pp. 161–557.

Lyell, C., 1997, *Principles of Geology*, Penguin (UK), abridged version of three volumes originally published between 1830 and 1833.

MacFarland, C.G., 1972, 'Goliaths of the Galápagos', *National Geographic*, 142: 5, pp. 632–49.

Mellersh, H.E.L., 1968, *Fitzroy of the Beagle*, Mason & Lipscomb (UK).

Merlen, G., 1999, *Restoring the Tortoise Dynasty*, Charles Darwin Foundation (Ecuador).

Palkovacs, E.P., Gerlach, J. and Caccone, A., 2002, 'The evolutionary origin of Indian Ocean tortoises (*Dipsochelys*)', *Molecular Phylogenetics and Evolution*, 24, pp. 216–27.

Palkovacs, E.P., Marschner, M., Ciofi, C., Gerlach, J. and Caccone, A., 2003, 'Are the native giant tortoises from the Seychelles really extinct? A genetic perspective based on mtDNA and microsatellite data', *Molecular Ecology*, 12, pp. 1403–13.

Porter, D., 1822, *Journal of a Cruise*, reprinted in 1986 by Naval Institute Press (USA).

Pritchard, P.C.H., 1996, *The Galápagos Tortoises: Nomenclatural and Survival Status*, Chelonian Research Foundation (USA).

Quammen, D., 1996, *The Song of the Dodo*, Hutchinson (UK).

Rothschild, M., 1983, *Dear Lord Rothschild*, Hutchinson (London)

Rothschild, W., 1896, 'Further notes on gigantic land tortoises', *Novitates Zoologicae*, 3: 2, pp. 85–91.

Rothschild, W., 1915, 'The giant land tortoises of the Galápagos Islands in the Tring Museum', *Novitates Zoologicae*, 22: 3, pp. 403–17.

Schweigger, A.F., 1812, *Prodromi monographiæ Cheloniorum sectio prima*, Königsberg (Germany).

Stoddart, D.R., 1967, 'Ecology of Aldabra atoll, Indian Ocean', *Atoll Research Bulletin*, 118, pp. 1–141.

Stoddart, D.R., 1968, 'The Aldabra affair', *Biological Conservation*, 1:1, pp. 63–9.

Stoddart, D.R. and Peake, J.F., 1979, 'Historical records of Indian Ocean giant tortoise populations', *Philosophical Transactions of the Royal Society of London*, Series B, 286, pp. 147–61.

Stoddart, D.R. and Wright, C.A., 1967, 'Ecology of Aldabra atoll', *Nature*, 213, pp. 1174–7.

Sulloway, F.J., 1982a, 'Darwin and his finches: evolution of a legend', *Journal of the History of Biology*, 15: 1, pp. 1–53.

Sulloway, F.J., 1982b, 'Darwin's conversion: the *Beagle* voyage and its aftermath', *Journal of the History of Biology*, 15: 3, pp. 325–96.

Swarth, H., 1931, 'The avifauna of the Galápagos Islands', *Occasional Papers of the California Academy of Sciences*, vol. 18, pp. 5–299.

Thomson, S.A., Irwin, S. and Irwin, T., 1996, 'Harriet, la tortuga de Galápagos', *Reptilia*, 2(4), pp. 46–9.

Townsend, C.H., 1925, 'The Galápagos tortoises in their relation to the whaling industry', *Zoologica*, 4: 3, pp. 55–135.

Townsend, C.H., 1931, 'The Astor expedition to the Galápagos', *Bulletin of the New York Zoological Society*, 33, pp. 134–56.

Townsend, C.H., 1942, 'Can we save the giant tortoise?', *Bulletin of the New York Zoological Society*, 55: 5, pp. 109–11.

van Denburgh, J., 1914, 'The gigantic land tortoises of the Galápagos archipelago', *Proceedings of the California Academy of Sciences*, Series 4, 2:1, pp. 203–374.

Index

Abbott, William, 228
Abingdon Island *see* Pinta
Agassiz, Alexander, 142
Albemarle Island *see* Isabela
Aldabra, 21–2, 27–31, 43, 79–84, 106,
    121–9, 134, 137, 138–9, 206,
    227–39, 251, 252–5, 256–61,
    264–9, 272–3
    Campaign to save tortoises, 122–4,
      144–53
    Commercial exploitation of,
      144–53, 228–30
    Estimate of tortoise numbers,
      149–51, 228–30, 237–8
    Military base and, 230–7
    Plans to relocate tortoises, 146–8,
      151–3, 227–8
    Scientific study of, 230–9
Army Museum, London, 201–2
Arnold, Nicholas, 252–5, 257, 258–9,
    266
Austin, Jeremy, 258, 266
Australia Zoo, 178–81, 187–9

Bainbridge, Commodore, 98
Barrington Island *see* Santa Fe
Baur, George, 142
Beck, Rollo, 213–19, 222–3, 225
Bell, Thomas, 49, 50, 56, 62, 63
Bibron, Gabriel, 63–5, 120–1
Bour, Roger, 252–3, 257
Bouton, Louis, 117, 118
Buckland, William, 48, 66

Bullock, Otis, 160
British Museum, 51, 58, 60, 74, 118,
    144, 191, 193, 195, 199; *see also*
    Natural History Museum,
    London

California Academy of Science,
    218–24
Canning, Laura, 253–5, 256–7
Cape Verde Islands, 56–7
Charles Darwin Foundation, 241–8,
    271–2
Charles Island *see* Santa María
Chatham Island *see* San Cristóbal
Clark, George, 115–16
Cocos Islands, 90, 164, 219
Colnett, James, 87–92, 93
Comores Islands, 268
Cook, James, 88, 176
Cornell, James, 160
Covington, Syms, 43, 51, 60, 190,
    191, 193–4, 195, 199, 200, 201,
    204
Cowley, Ambrose, 15–16
Curieuse Island, 227–8

Dalyell, Tam, 235
Dampier, William, 15–18, 85, 89
Darwin, Charles, 31–71, 100, 109–10,
    118, 136, 137, 214, 219, 264
    Beagle and, 32–44, 177–205
    Encounter with tortoises, 40–2
    Günther, Albert and, 199–201

Picture Acknowledgements

The author and publisher would like to thank the following for permission to reproduce illustrations.

© The Natural History Museum, London: Page 165, Page 119. Courtesy of the Wildlife Conservation Society: Page 95. © Godfrey Merlen: Page 243. Down House, Downe, Kent, UK/Bridgeman Art Library, London: Page 37. © Rollo Beck, Special Collections, California Academy of Sciences: Page 216. Courtesy of the Illustrated London News Picture Library: Page 19, Page 208. The Rothschild Archive, London: Page 155.

DATE DUE

JUN 2 8 2006		
JUL 0 5 RECD JUN 29 2005		
GAYLORD		PRINTED IN U.S.A.

SCI QL 666 .C584 C46 2006

Chambers, Paul, 1968-

A sheltered life